Conversation and Community

The Social Web for Documentation

2^{**nd**} **Edition**

Anne Gentle

xmlpress.net

Conversation and Community
The Social Web for Documentation
Copyright © 2009-2012 Anne Gentle

Credits

Illustrations:	Patrick Davison
Author portrait:	Beverly Demafiles Schulze
Top photo on cover:	VM-housing by seier+seier on Flickr
Bottom photo on cover:	MacBook Pro Keyboard by Kolin Toney on Flickr

Disclaimer

The information in this book is provided on an "as is" basis, without warranty. While every precaution has been taken in the preparation of this book, the author and XML Press shall have neither liability nor responsibility to any person or entity with respect to any loss or damages arising from the information contained in this book.

This book contains links to third-party web sites that are not under the control of the author or XML Press. Inclusion of a link in this book does not imply that the author or XML Press endorses or accepts any responsibility for that site's content.

Trademarks

XML Press and the XML Press logo are trademarks of XML Press. All terms mentioned in this book that are known to be trademarks or service marks have been capitalized as appropriate. Use of a term in this book should not be regarded as affecting the validity of any trademark or service mark.

XML Press
Laguna Hills, CA
http://xmlpress.net

2nd Edition
ISBN: 978-1-937434-10-6

Table of Contents

Foreword – First Edition

A few years ago this book could not have been written, because the phenomena it describes were just poking their heads out of the sea, and no one could predict what form their evolution would take. A few years from now this book will be unnecessary, because we'll all be participating so fully in the phenomena that newcomers will take to them like ducks to water. You are fortunate to have this book at this moment, for you can lead the next generation of information providers into the era of expert/amateur interaction.

Anne Gentle purports in this book to give you the tools and insights to create growth-oriented educational experiences for your clients in this age of collaborative learning. To bring home the goods, Anne has combined hard-won insights from every angle: academic research, reports from the field by other cutting-edge practitioners, years of experience in corporate technical writing departments, and personal practice as a contributor and organizer in some of the most sophisticated of open-source, community-oriented authoring sites.

I've exchanged mail with Anne and worked with her on FLOSS Manuals, an experiment in free documentation that takes center stage in many of the examples in this book. Having toured some of the same ground as Anne, having enjoyed some of the same views from the commanding heights, and having been caught in some of the same downpours, I can attest that she has the landscape right.

FLOSS Manuals is just one of the many projects on the Internet that show the power of people working together to educate the public with documents open to all for editing and distribution. Among the lessons these projects offer is the tremendous energy that arises from public engagement. But they also show that this energy is more potent when managed well.

Because client education, as a key element of client satisfaction, is the concern of anyone working on a project with outside stakeholders, this book should claim a wide audience. The main targets of the book are technical writers at companies experimenting with social media (also known as Web 2.0). But among the barriers to collaboration sliced away by Anne's scalpel is the distinction between official technical documentation, marketing-oriented white papers, release notes, customer service forums, and even customer comments in the field. If you work in technical development, marketing, customer support, or that most intriguing new job description – community manager – this book has news for you.

Every source of information with which we're familiar – journalism, education, government, and certainly technical documentation – is abandoning the oracular view of information. In simpler days, those of us in documentation and publishing found experts and simply proclaimed their insights to the masses. Now we realize that many of those insights originated among the masses in the first place. Furthermore, writers and publishers find that their work doesn't get read unless the expert conducts a dialog with the audience.

We may know what they need to know, but only they can tell us what they need to know.

I'm hoping that the context for the previous, artfully crafted sentence ensured it had the proper impact, but in case your reader response didn't work as I had hoped, let me spell it out. Experts possess knowledge that other people wish to mine, and often will pay for. But the experts often can't anticipate the questions that non-experts will have. The experts are shocked to discover the areas where non-experts experience difficulties. Once they understand their audience better, most experts radically alter the topics they discuss and how they present them. Thus the importance of conversation and collaboration between experts and non-experts.

Many of us worry about our future in an environment of blogs, wikis, and stakeholder forums, where anyone who knows anything can share it. The question "How can I earn a living?" must be approached in *Through the Looking Glass* fashion, by walking away. Join the online conversation. If there's a place for you as a professional, you'll find it there.

Occasionally I talk to someone who believes that self-help and self-organization are taking the whole pie and that the best sugarplums will pop up on their own. If that were true, the world wouldn't need professional editors and information architects, and it wouldn't need Anne's book. But these misguided futurists have it profoundly wrong. Experts and non-experts need to work together.

I finally got provoked when one programmer boasted to me that he never read any documentation. He looked at the source code of the software he was planning to use as a start. He would write code 'til he hit a barrier, then ask a question online, then go back and write code 'til he hit another barrier, and so on.

I responded to him, "You would never design a software system using such an undisciplined and ad hoc methodology. Why do you tolerate your learning experience to be designed in that manner?"

The programmer didn't appreciate that each person must bring some expertise to benefit from experimentation and collaborative learning. Modern technical training, with its heavy emphasis on both experimentation and collaboration, draws heavily on John Dewey's classic theories of learning as a dialectical process between the learner and her environment – the more prepared the learner is, the greater heights she can attain.

You probably remember rejecting a book as boring and irrelevant, only to come back and embrace it later after a phase of personal growth. Collaborative forums can foster this upward spiral. The techniques in this book help you not only improve the forums, and use them to improve your documentation, but also use the forums to prepare readers to benefit from the documentation.

So, what can you offer as a writer or other professional to foster this upward spiral and give readers the best available educational experience?

You can create tools and forums

Most readers stick to familiar ways of communicating. The hoary old mailing list format, and even newer documentation models such as wikis, fail to exploit the power of user contributions. It takes work to create a system that efficiently accepts, formats, and disseminates information provided by public participants. And those are just the start. Popularity ratings, tagging mechanisms, and sophisticated search options can dramatically increase documentation's usefulness as well. Take the lead to develop and train your community to use a powerful interactive educational system, and their productivity will soar. You'll probably generate more interest in your professional contributions along the way.

You can edit and train contributors

Everybody can use an editor. Routine proofreading for grammar and consistency aren't usually important for casual online contributions, but anything longer than a couple of paragraphs could benefit from a look at its purpose, structure, and information gaps. A small investment of professional time can make the document much more useful. Remember that you are a collaborator, so you need to work dialectically with the author to find the best approach to the document. And you'll both learn something from the process.

You can help people find information

A big part of writing is going back to straighten out a document. Because online documents tend toward drastic fragmentation and reflect many different voices, the role of a professional writer and editor faces a steeper challenge but a concomitantly larger benefit to the overall site. You can spend time organizing documents and creating pathways through them, creating links, and putting up portal pages. Don't feel shy about intervening directly in the documents themselves, by adding introductions and transitions and changing headings to indicate more accurately what's inside the document.

You can recruit, cheerlead, and inspire

Most people have something to contribute, but they wait to be asked. Once asked, they're more than happy to help out. You need to determine where the needs are and who can fulfill them. You also need to offer support for them to create a document of high quality and promote the result so they feel their work is rewarded by exposure.

You can consolidate the best material into a professional document

Your own documentation should be the flagship for the community. You can use community input to improve it and add some research of your own to take it to the next higher level of community trust.

Reading Anne's book brought me a number of satisfying moments. She quite properly recognizes the importance of both documentation and search: the notion that both availability and findability are fundamental documentation requirements. I'm glad she attacked the stance that we serve our project best by always presenting it in a positive light. She understands the role of audio and video media in the current educational landscape. And she even draws insight from one of my favorite and chronically under-appreciated research projects, John Carroll's Minimalist documentation.

The book is not as tightly organized or carefully paced as an ideal expository text would be. This is because Anne offers so much, has so few sources to point to when making her core points, and is covering fast-changing areas whose participants offer new insights literally every week. She manages to pull it all together.

I think the author of a foreword is expected to say he "couldn't put the book down 'til it was done." I'm afraid I can't offer Anne the opportunity to make that boast. In fact, I was constantly pulled away from her book by her references to fascinating documents that added extra fodder to her argument or provided new perspectives. I'm sure Anne's web site will host equally enticing content. Let this book be an anchor for your exploration of collaborative document production, so that your own documents can be anchors within your clients' learning ecosystem.

Andy Oram
Editor, O'Reilly Media (organization for identifying purposes only)
12 May 2009

Preface

This book grew out of my experimentation with open source, blogging, writing in a wiki for online help, connecting with community members, and a compulsion for writing things down. I was also prodded somehow by my complete adoration of Google search, which I felt would change forever the way that users find information that helps them complete a task, even one as simple as slicing a tomato[1] or folding fitted sheets.[2] Everyday tasks are documented on the web by everyday people. The start of this people-centric revolution has been described as the second generation of the Web. Web 1.0 was about data and display, but Web 2.0 merges data and display with user-centric design and ideas.

While people debate the cult of the amateur[16] and spread uncertainty and doubt, professional writers now have the tools to collaborate with their audience easily for the first time in history. How we seize this opportunity and how our audience responds and becomes a part of this revolution will determine our success in this new environment.

Nearly all of the discussion in this book surrounds new ideas for documentation, whether you are writing documentation as a film maker, corporate marketer, technical writer, programmer, or manager. Sometimes these "new" ideas uncover age-old truths about communities and people's behavior and habits. Sometimes you have to experiment on your own time with your own dollars and tools to prove that a technique is worth an investment. That try-and-see attitude is what the cutting edge is all about.

---

[1] http://www.ehow.com/how_1682_slice-tomato.html

[2] http://www.ehow.com/how_6067_fold-fitted-sheet.html

Beth Kanter, a trainer who teaches non-profit organizations how to bring social networking to their business, compares experiencing social media for the first time to a first-time sexual experience – you can't describe how it feels until you experience it. In the same way, you can't describe or measure the value of a technique or methodology until you try it and analyze your results.

Many people would argue, "how can you find the time?" and would describe social media dabbling and experimentation as a waste of time. Clay Shirky, author of *Here Comes Everybody: The Power of Organizing Without Organizations*[31] and *Cognitive Surplus: Creativity and Generosity in a Connected Age*[32], offers a counter-argument pointing out that the 200 billion hours per year spent watching TV in the US pales in comparison with the 100 million hours spent to create Wikipedia (see "Cognitive Surplus Visualized"[69], by David McCandless, for a visualization).

By finding and making the time for this experimental discovery, you may save time and effort and increase the quality and perceived value of your content. Plus, you may find interacting with others more fulfilling than the one-way communication offered by your television or other media outlets.

What's new in this edition

In the three plus years that have passed since I completed the first edition, I have continued to learn from my experiences as a community leader, content strategist, technical writer, documentation automation and system analyst, and student of web analytics. I have also met people who have generously taught me and offered their lessons learned. In this edition, I have tried to include as many of their stories as possible through interviews.

I have added deep dives into my areas of interest: content strategy, web analytics, and open source documentation. And I have made revisions based on input from some of the many university students who have used this book as a text book.

What's in this book

Use this book to help you experiment with social media, social networking, and social relevance, and to analyze and interpret your results.

It offers descriptions and definitions for the technologies and publishing methods that make up this new way of thinking about content, and it provides ideas for defining your role as a content author or provider. You will find planning and implementation suggestions and advice as well as considerations for choosing your role and goals as a writer or provider of content.

This book includes a chapter on measuring the effectiveness of these new techniques and proving their value to various stakeholders. And you will find specific ideas for integrating conversation, community, and collaboration into documentation.

In this book you will learn about enabling conversation and community in your documentation using social media and social networking. Our world is shifting, and the definition and scope of documentation is moving along disruptive fault lines. Mark Baker describes this in a recent blog entry:

 But on the web, something new is emerging: communication that has the individuality and personal touch of a conversation, but the persistence and public availability of a publication.
—Mark Baker, "I am a content strategist"[43]

This book shares ways that we can manage this intersection of publication and conversation and work successfully with collaborators and their contributions.

What does it mean to enable a conversation, and how do you assess Web 2.0 tools and strategies such as wikis and blogs? In the landscape of technology products, the consumption of technical topics is often reduced to finding the right answer quickly, solving the problem, and moving on. End-users do not necessarily care about the source of the information or whether it was written by a professional; they judge the information solely on its ability to solve their problem.

So if you are professional writer, how do you fit into a landscape in which content must be constantly available and up-to-date and where blog entries get more visitors than your help pages? How do you ensure that your content has the same or better value than content from a myriad of online sources?

If you are a developer who wants to ensure that users have a good experience and get the answers they need, how do you respond to questions naturally or even presciently? If you are a content creator, such as a filmmaker or game designer, how do you ensure that you are entering the conversation and enabling community in your communications?

To answer these questions, and to help writers determine which social networking tools might help them communicate technical information to their end-users, this book examines the categories of social media and networking tools and provides pointers for evaluating each newcomer or old standby.

Related information

In the spirit of free and open sharing, links to all of the websites mentioned in this book are collected on delicious.com. Participate in building this book's future by adding new, relevant URLs on delicious.com using the tags *conversation*[4] or *community*[5] and adding me to your network.

About Anne Gentle

I currently work as the fanatical technical writer and community documentation coordinator at Rackspace for OpenStack, an open source cloud computing project. Prior to joining OpenStack, I worked as a community publishing consultant, providing strategic direction for professional writers who want to produce online content with wikis and user-generated articles and comments. I write a professional blog about writing, wikis, and information design at JustWriteClick.com.

I became interested in using wikis for documentation and decided that a hands-on apprenticeship would be the most efficient way to learn about wikis. I have been volunteering for the One Laptop per Child project, writing end-user documentation for children, parents, and teachers across the world, using open source software that could change the way education happens in under-developed and under-served nations.

FLOSS Manuals, a toolset and community dedicated to writing free documentation for free software, shaped many of my experiences with community documentation. I have some history on the web now as I started blogging in 2005 for BMC Software.[7] I would not

---

[4] http://del.icio.us/annegentle/conversation
[5] http://del.icio.us/annegentle/community
[7] http://talk.bmc.com

have the amazing opportunity I have today to put into practice these ideas each and every day if it weren't for the way this book has shaped my career path, my passions, and my love of community.

Acknowledgments

For a few years, I worked a 30-hour week, which allowed me to spend my time pursuing interesting projects and also work a longer "mom shift" in the afternoons and evenings. Without nap time, early bedtime routines for my kids, truly-caring child care, and above all, a wonderful husband, I never could have written this book, nor would I have learned the lessons that enabled me to write it.

I offer special thanks and gratitude to my husband Paul for encouraging me. His involvement in the distributed.net community brought us to Austin, Texas, and showed me the power of volunteers sharing a common cause.

My good friend Kelly Holcomb skillfully edited the initial drafts of the first edition of this book. She read it, edited it, and asked wise questions in the margins, which compelled me to answer them.

I owe a huge debt for the education I have received from FLOSS Manuals founder Adam Hyde and SugarLabs coordinator David Farning. They both read extremely early drafts of this book and encouraged me throughout the process. Adam generously contributed most of the content about Book Sprints. Working with him has been inspirational. He has assembled a great crew at FLOSS Manuals, which connected me with Patrick Davison, who did the interior and cover designs for the first edition, and whose illustrations are still part of the second edition. And without FLOSS Manuals I might not have met Andy Oram, who graciously read the book, offered guidance, and wrote an insightful foreword for the first edition.

The cover photographs were taken by "seier+seier" and "Kolin Toney." "krossbow" and "Pathfinder Linden" provided the Second Life images in Appendix A. Thanks to all four of them for making their photographs available on Flickr[9] under a Creative Commons License. Thanks to "nolnet" for giving me permission to use photos of his Lego refrigerator in Chapter 4.

Thanks to Lana Brindley, Lisa Dyer, Dee Elling, Sarah Maddox, Shaun McCance, Eric Shepperd, Victor Solano, and Janet Swisher for permission to include their interviews and for sharing all their lessons learned.

Thanks to Eve Smith and Easter Seals for giving me permission to reprint the Easter Seals Internet Public Discourse Policy in Appendix B. Thanks to Frank Gilbane and Outsell, Inc. for giving me permission to reprint the case study in Appendix C. Thanks to Scott Abel for giving me permission to reprint his interview with Victor Solano in Chapter 7.

Scott Abel, Sarah O'Keefe, Alan Porter, and Will Sansbury read a hack of a draft and offered wonderful insights from their varied perspectives. I learned from each of them. Without their generous gift of knowledge, wisdom, experience, and time, this book would not be as useful as I hope it will be to you.

---

[9] http://flickr.com

1

Towards the Future of Documentation

Tell me... and I will forget.
Show me... and I will remember.
Involve me... and I will understand.

—Confucius

As writers and content creators we are witnesses to the age of information shifting to the age of interaction. One week of content from the New York Times contains as much information as a person could get in a lifetime in the 18th century, according to information architect Richard Saul Wurman in his 1989 book, *Information Anxiety*[36]. The speed at which we access this information is constantly improving – with fiber optic technology you can receive an amount of information equal to 10 full CDs per second. Filling a CD with that much text would take twelve years typing 100 words per minute! The only way people can handle this much information is to build relationships and act as filters for each other.

Ironically, some content creators do not yet see a link between *online help* and the blogs, wikis, and forum posts with which users are finding *help online*. I have seen in writers' email lists such quotes as "Sorry, I absolutely refuse to discuss anything, even cats, on a blog!" or "I'd rather not trust an encyclopedia that would accept me as a contributor." Even if the Groucho Marx reference was meant in jest, how can we learn from the age-old community behaviors that are now being applied to documentation?

Documentation as conversation means getting closer to users and helping them perform well. User-centered design has been touted as one of the most important ideas developed in the last twenty years of workplace writing. Web 2.0 gives users a chance to interact with information and other users.

Now, writers can take the idea of user-centered design a step further by starting conversations with users and enabling user assistance in interactions. Writers have more conversation-starting tools at their disposal than at any other time in history. Tools may include blogs, wikis, forums, and social networking sites, but may also involve photos, simple stick figure illustrations, videos, virtual worlds, or instant messaging.

How people communicate about technical topics today

The harder it is to get something to work, the more comforting it is to get help from a real person. The increased complexity and reach of technology has driven us to reach out to people through various methods to help us understand and use that technology.

How has changing technology affected how these conversations are held? An example of how technical documentation has changed over the years is sewing patterns. Historically, people talked face-to-face to teach others how, for example, to sew a quilt for a bed cover. Later, the patterns were written down, and with the advent of printing, patterns could be produced for anyone to use at home. Today's quilting instructions are disseminated on the Internet, mailing lists, forums, and blogs,[1] and online communities are forming around the passion for quilts and quilting.

Over the years, the cost of conversation has dropped. We have gone from the need to design, write, and publish (a long, sometimes arduous process that was out of reach for ordinary people) to print-on-demand services and other "instant" methods for publication. With microblogging services like Twitter, you can merely have a thought, decide to express it, type it on your smartphone, and immediately push it out to all your followers. It's a searchable piece of content and can even be reused as a sidebar on a "macro" blog. Communication is immediate and filtered as the recipient requests.

Using questions and answers

As a professional writer entering the wiki world as an apprentice, I had to determine my goals for designing documentation that would be part of a conversation within a community. I learned that conversations are typically measured in three areas within an enterprise: public relations, customer support, and sales and marketing.

With this framework in mind, when I volunteered to write documentation for parents, teachers, and students using the One Laptop

---

[1] My mom writes a quilting blog at quilterjan.blogspot.com and has over 60,000 hits since she started it in 2007.

per Child hardware and open source education platform, I determined that my best fit was in the customer support area.

When you start supporting customers, especially for technical information, you want to analyze their tasks and anticipate their questions. In *The 4-Hour Workweek*[9], Timothy Ferriss says that the first web page you create when automating the business of selling a product or service should be *a list of actual asked questions*. He says you should build up your list of questions over time as you answer the questions yourself.

This type of do-it-yourself approach is excellent for producing documentation like the horse-training books by Kathy Sierra. Kathy Sierra is a master trainer, creator of the HeadFirst series, and author of the Creating Passionate Users blog.[2] According to her, books that embed questions right within the instructions are "the best user manuals EVER." Her blog entry titled, "How to get users to RT-FM"[81], states that a good manual for a complex product should include at least five distinct sections: Reference Guide, Tutorial, Learning/Understanding, Cookbook/Recipe, and Start Here.

Her detailed descriptions of each section describe, in essence, John Carroll's basic concepts of minimal documentation.[3] And the principles of minimal documentation play a major role in creating documentation that fits into user conversation. Minimalist principles encourage both writer and instructor to:

- Get the user up and running quickly
- Let the user think and improvise
- Focus on real work and real goals
- Make use of the user's prior knowledge
- Use error recognition and error recovery as learning helpers

---

[2] http://headrush.typepad.com/creating_passionate_users/
[3] Minimalism was first described by John Carroll as an instructional design philosophy at the IBM Watson Research Center in the 1980s and presented in his book *The Nurnberg Funnel*[6].

Kathy Sierra seems to be describing manuals that support popular consumer products, like TiVo digital video recorders or MO-TORAZR mobile phones. Learning about consumer products may inspire more passion and excitement in writers, both amateur and professional, than writing about products that are used in the workplace (and perhaps for boring or dirty jobs). While some may be passionate and inspired by open source products, and others may enjoy documenting medical devices, for the most part, the documents that writers are paid to write are for not-so-exciting products that help people do their jobs.

So, although Sierra's outline does not seem widely applicable for all technical documents, it makes one wonder just how these types of sections could be worked into the existing documents. Would any particular audience appreciate sections that have the type of information that she's talking about? Most likely the answer is yes. And wouldn't we all like to be a little more passionate about the tools we use every day?

In addition to finding passion in the tasks users do, writers should be applying customer support data to help them keep up with changes in questions as time goes on and as releases go out the door. Much like the most popular Google search hits, the most frequently accessed questions and answers should rise to the top of a customer support knowledge base.

You can also gain valuable insight into the most important and urgent questions by interviewing the customer service and support representatives who spend most of their time on the phone or in the field with customers. Ann Rockley advocates this analysis as a first step in developing a unified content strategy. Her approach has helped customer service representatives be more efficient and effective through better content, search, and navigation.

Meeting shifting expectations

In developed countries, technology is pervasive in all parts of life – home life, school life, and work life. As a result, expectations for technical documentation are shifting. Prior to the Internet, people would carefully store printed manuals on a bookshelf for reference. Today, people use a search engine as their starting point for troubleshooting a device. For example, I still saw click-throughs from people searching for "BMC Performance Manager" nearly two years after posting blog entries about the BMC Software product "Performance Manager."

The booklet that comes with a consumer device may still be the best place to look for assistance, especially because of the multiple translations available for international audiences, yet a visual, video-loving user may search on YouTube for troubleshooting advice. And others might just use the interface itself until they figure it out. Still others might consider a friend or co-worker the best source of technical expertise and rarely use the documentation directly (although the expert friends or co-workers are certainly intimately familiar with their favorite sources of technical information).

Using search

Search technology, and its application by users, has changed the face and entry point to technical documentation. The "entry point" is the page your user lands on first, also called a "landing page." Sometimes landing pages are text-heavy and feature-focused rather than task-oriented. Users are more likely to enter a couple of keywords into a search form as their the starting point for reading online help.

How many times have you heard someone say, "I found the following Microsoft Knowledge Base article using a Google search" (or have said the same thing yourself)? Microsoft has provided excellent content in their Knowledge Base, but a competitor's search engine often brings users to the site. Although the Knowledge Base must have a search engine available to visitors (much like a manual of some sort is still required for many consumer products), when a visitor finds the content by using another search engine, all the resources, time, and money spent on providing a search engine on that Knowledge Base were wasted. The company experiences zero return on investment for the search engine, but total return on investment on the content itself.

Search relevance has to do with Google's secret algorithms and the calculations used for ranking by any search engine. This relevance brings a user to your content when they are seeking an answer to a specific question. If your help system does not follow Google's basic Webmaster Guidelines,[4] it may not be found as often as blog entries on the same topic. For example, Google recommends that you use fewer than 100 links from a sitemap, and warns that not every search engine spider will search dynamic pages that have multiple parameters in the URL.

You may have noticed that a recently updated blog entry or wiki page tends to have a higher position on the Google results page than less recent pages. A comment[5] on an entry in the Britannica blog makes this point nicely: "Perhaps another idea would be to ask Google WHY the social media (i.e. Wikipedia) entry for a topic ranks so much higher than the vetted, Britannica entry?"

Secret algorithms or not, as a writer you need to understand your user's entry points. Search is important to conversational documentation because of the importance search engines place on frequently-updated content. If your user assistance system is offline or infrequently updated, your users may not find answers by searching with popular search engines. For example, Google gives a higher search ranking to more frequently updated content.

---

[4] http://google.com/support/webmasters/bin/answer.py?hl=en&answer=35769
[5] http://www.britannica.com/blogs/2008/11/the-fast-food-information-age-we-are-what-we-read/

Naturally, you know your audience best. If they do not require online access to use your product, they may not search for answers to questions in a search engine.

Merely using a blog or wiki as a holding place for content may improve the search ranking. You may think that your users will search within the help in the product itself, but you may find, as the Britannica blog writer did, that between 90% and 98% of library users in 2008 assumed they could get all the relevant information on a subject by simply searching on Google.

If these results are also true for users of the product that you document, then you need to try to get search words pointing to your online user assistance. A good first step is to ensure that your online user assistance is available on the Internet to be found by search engines. Also find out what your users search for, and then determine how you want to improve the ranking of those searches. Perhaps you want to integrate user-generated content into your user assistance if search engines tend to value those entries more than others.

You can also place a carefully-crafted search index as the entry point to community content. Adobe created a Google Custom Search Engine as a front-end to community content and hand-selected the most useful community offerings such as blogs and video tutorials at community.adobe.com. By indexing only the highest-quality outside resources, Adobe helps build trust with the readers that use the search hits to find information.

The changing roles of writers

In a February 2008 interview with Tom Johnson in his "Tech Writer Voices" series, STC President Linda Oestriech said, "We aren't the technical writers of the 70s."[7] I might add that we aren't even the technical writers of the 90s. How have changing technology and expectations influenced technical communication and writing today compared to 10 or 30 years ago?

Many have been wondering about the state of technical communication and looking to history for insight. For example, in *Where Did Technical Writing Go?*[82], Jared Spool, a usability consultant, describes a past when technical people and engineers expected to learn every detail about a product and had no expectations of simply being able to pick something up and know how to use it. He describes how the user manuals weighed more than the product itself (and this was when computers filled a room). In those days, he says users invested enough in the technology that they wanted to learn the product and started with reading the manuals.

In contrast, Kathy Sierra describes a successful manual today as a user-friendly page-turner that users learn from – one that pushes users to higher levels of understanding by being motivational as well informational. Technical writers can deliver that user manual in the right corporate environment, or publishers can hire writers to deliver that user manual for sale in bookstores.

Technical communication is not just about writing; it has always spanned multiple areas of expertise. User interface design and usability play a large part in communicating technical information so that users can do neat things with a technical product. Because our audiences use technology more often, but also demand simplicity and elegance, many products have become easier and more intuitive to use. As more products are available in a global marketplace, words are increasingly replaced by images and illustrations. Video editors and video producers are technical communicators as they develop training courses. The market for words and printed books still exists,

---

[7] http://idratherbewriting.com/2008/02/05/interview-with-stc-president-linda-oestreich-the-direction-the-stc-is-heading/

but favors "missing manuals," guides for "dummies," and other books that teach or instruct.

Defining conversation

Conversation as a discourse is nearly undefinable since it spans so many speech events and environments, from informal to formal, intimate to public, and argumentative to collaborative, plus many more. But some communication contexts are useful for documentation. Nicky Bleiel provides one example in her Content Wrangler article, "Convergence Technical Communication: Strategies for Incorporating Web 2.0"[46], where she describes communities of practice that allow users to participate in conversations.

Imagine that you were assigned to answer customer support calls at a mobile phone network supplier for a day. You would establish a connection with each customer, associate a name with a voice, face, or email address, and then from a quick customer database search, you would know who the customer works for and what that customer's role is. Thanks to your company's call log, you might know the exact problem each customer is wrestling with and be able to pinpoint the information needed to solve that problem.

If writers had that much information every time they started writing, they could be more effective for one individual customer. But technical guides usually need to help hundreds of customers. So, imagine instead that you are reading a deep technical debate in the discussion area of a wiki, on a customer's blog, or a 140-character Twitter post. How would you rewrite your documentation to answer that user's exact question and then make the information accessible to any user? And can you step into the conversation by posting a response that gives that user the right information at the right time?

Asynchronous conversations

Nearly all documentation can be considered an asynchronous conversation between reader and writer, meaning that the messages can be written or received at any time regardless of the sender or receiver's schedule. The reader is an audience that the writer has to consider before determining what to write. As Ginny Redish says in her book, *Letting Go of the Words: Writing Web Content that Works*[27], the writer's job is "helping people get just what they need, when they need it, in the amount they need, as quickly as possible." Good information chunking and organizing practices create a just-in-time conversation to help someone accomplish his or her goals.

Levels of conversation

Adam Hyde, founder of FLOSS Manuals, finds that conversation has a clear and direct value in the in-person events called "book sprints" that create content from outlines in a short period of time, such as a week. He perceives that the conversation takes place on at least four levels:

■ **In the room:** With all participants seated around the same table, conversations need to be fostered. Unlike traditional writing processes, this is a noisy environment and chatter should be encouraged. It is through these across-the-table conversations that participants harmonize their ideas about what they are writing.

■ **In the text:** The texts themselves are conversations. Do not be protective about the content you contribute – encourage others to read it, change it, delete it, etc. Having many eyes continually combing through the content is a fundamental tenet of Free Software and serves equally well in this environment. This constant change is itself a conversation happening in the text, and it is essential for improving and harmonizing the content.

■ **Outside the room:** Be in touch with people outside the room and ask for their take on the text. It is important to know if the text is going in a direction that can be understood by people who have not been privy to the conversations in the book sprint room. Encourage remote participants to provide feedback via Internet Relay Chat (IRC), email, editing, and the FLOSS Manuals web chat.

■ **With the reader:** When writing, use a direct and friendly and conversational voice. One trick to getting disk jockeys to relax and sound "normal" on air is to cut out a picture of a face, stick it to the microphone, and ask the DJ to talk "to" that person. This works. We have not tried this approach yet in book sprints, however the same effect can be achieved by thinking of the reader as someone you know well and respect. Write to that person as if you were explaining something you know he or she doesn't understand.

Agile development

Agile development techniques may involve writing user stories that can be developed and designed within a "sprint" or iteration that may be as short as two weeks. Teams applying this type of development strive to develop robust software rapidly – including testing and documenting the product – so that the new story can be played out immediately after the end of the iteration. By leveraging the Agile user story – which should be user-centric and task-oriented – writers can put themselves into the place of a customer. Because Agile development encourages developers to write the user story while developing the product, the goal is to match the user's expectations quickly. In at least one important way, Agile development lends itself well to Web 2.0 techniques by pushing the use of user stories earlier – towards the development phase rather than the post-release phase.

If we can get closer to our users by discovering their documentation needs, we can better understand their product needs. Then, we can help build solutions to these needs during the design phase of our products, rather than tacking on documentation as an afterthought. Collaboration with users can help us improve collaboration in the workplace during product design.

You can read more about Agile development techniques and the Agile manifesto at http://-agilemanifesto.org.

Why move content to the social web?

As Harry Miller at Microsoft pondered in his mid-2000 podcast, "The IM Model of Tech Writing"[70], what if a user guide had to read like an Instant Messaging, or IM, conversation – quick, real-time questions and fast answers, tailored to nearly every potential customer situation?

With Twitter's popularity as a customer care platform growing rapidly (in March, 2011, Twitter grew by an average of 460,000 new users every day[38]), we may be seeing the birth of a help-on-demand user's guide. In fact, Mozilla uses a crowd-sourced Twitter support model to help people struggling with Firefox with their Army of Awesome.[8] If you love Firefox and want to help others, you can respond to tweets requesting support.

Different users have different expectations and needs for documentation. Some users expect precise answers. Some want to impress people with their knowledge and efficiency. Others might have a particular aversion to truly learning a product because they use the product only once every other month to perform a specific (perhaps

---

[8] http://support.mozilla.org/army-of-awesome

boring) task. The manual that you write might not respond to all of these needs just yet. But if you keep enabling conversations with users, your documentation can anticipate questions and give answers so responsively that your users might wonder if there's a ghost in the machine.

But even if your documentation can't "talk back" to your users, it can help users talk to each other and make connections that help them do their jobs well, play with technology at home, or learn something new in a classroom setting. Instead of concentrating on single sourcing methods or the tools of the trade, think about documentation and user assistance as a multi-channel communication device, perhaps with the help of some social technology applications.

Many skills that are valuable in delivering traditional user assistance are easily transferable to social network integration for user assistance. These skills include:

- Excellent communication
- Concise writing
- Good design
- Careful planning

Technical writers have plenty of good reasons to avoid actual conversations with customers. We are not usually trained to handle an angry customer or to troubleshoot the product at the technical level that is necessary. But technical writers are typically good at learning quickly and applying technology to solve problems.

You might think that social media or social networking is meant only for young people, or that only a person with too much spare time would appreciate things like social bookmarking or Second Life. But these technologies do not belong only to the young, the technically savvy, or the extremely extroverted. The demographics for both Facebook and Twitter users show trends towards an aging group. The second fastest growing demographic group for Facebook in January 2011 includes people over the age of 55[9] and more than 40% of Twitter users are age 35-49.

Technical communicators are already skilled in many of these technologies and just have to learn to apply them to communication with users. However, missteps may cause setbacks in your plans to enable customer connections with your documentation, so it's useful to seek out internal coaching in the area of customer communications. Also, you should make sure your plans mesh with any overall corporate social media strategy.

Social media, social networking, and now the social web

Social media is still being defined, but over the past few years, more people have weighed in on the definition. It's an interesting use of social media to study the definition of social media. For example, the Wikipedia entry about social media points to the popular tech blogger Robert Scoble's blog post about social media in which he defined social media as news media pitting old media against new media. But even though social media is compared to newspapers and journalism, technical writers can think of themselves as interviewers and journalists. A journalism background may help you see ways to present information in a news-delivery style. Scott Abel, creator of The Content Wrangler[10] website and community, has a journalism background and purposely runs The Content Wrangler site more like a magazine than a blog.

Many companies invest in social media projects through their marketing or advertisement departments. Technical writers, how-

---

[9] http://www.istrategylabs.com/2011/01/2011-facebook-demographics-and-statistics-including-federal-employees-and-gays-in-the-military/

[10] http://thecontentwrangler.com

ever, should not dismiss the application of social media to their deliverables just because social media appears at first glance to be more of a marketing function or a method for building up a brand.

Social media, social networking, and the social web are related concepts but separately defined from Web 2.0. As social media researcher danah boyd, (she prefers no capitalization of her name), pointed out in a talk at the Microsoft Research Tech Fest ("Social Media is Here to Stay... Now What?"[48]), Web 2.0 has different meanings for technologists, businesses, and users. She also says, "Social media is the latest buzzword in a long line of buzzwords."

No matter what the preferred term is, those of us researching social aspects of information dissemination can find many applications beyond news reporting and brand building for social media. The next chapter explores some of the basics and then dives into examples of web applications that enable social media.

2

Defining a Writer's Role with the Social Web

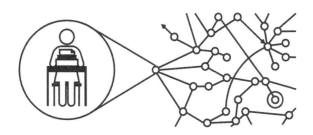

The more original a discovery, the more obvious it seems afterwards.

—Arthur Koestler

As a writer, you need to find the correct value proposition for your work and determine whether your role is an enabler of conversation or an instigator of conversation. In a world of rock star coders, you want to find out whether you are a sage on stage or a stage hand. Or perhaps your evaluation will determine that no social media action is the correct action.

You might already be aware of the challenges of introducing collaborative or community-driven content in an enterprise environment. Many people are skeptics, asking for proof of the efficacy of the methods. This chapter discusses strategies for determining your role and offers strategies for a phased approach to integrating the social web in your writing that can help answer the skeptics.

Challenges and opportunities

Once you get enthusiastic about a particular social networking tool or medium, you may find that you spend an inordinate amount of time on the site or reviewing the tools used for the site. Be aware of your priorities, and be cautious of making a tool appear to be a time-wasting, productivity-sucking black hole before others can give it a chance. Remember that your online presence records the amount of time and the time of day that you work with a tool. You can avoid too much of a good thing if you're smart about your time allocation. You might also need to participate in an online community after normal business hours to ensure that your company trusts you to do your "real" job first and experimentation second.

It's important to find a balance between allowing an individual's authentic voice to speak on behalf of an organization and the requirements of institutional messaging and brand preservation. Be sure to follow all related corporate policies and study the messaging that your company prefers to maintain.

It's also possible that you are ahead of the curve and need to help others see ways to apply social technologies for the company. You could get involved in a community related to personal interests outside of work to understand how social technologies can be applied before attempting to apply what you've learned at work. This approach works especially well if your organization is extra-conservative or if there is no demand for content with social connections yet for your company.

What about the ubiquitous argument, "Sorry, but we just don't have the resources for this type of project"? You might need to barter for resources or entice someone's time and skills off the clock. Think of the reasons that people are motivated to contribute to a project, such as recognition or respect, and see what you can offer in return. Also realize that if the resources truly are not available, but you think you can save time and money with social or community content, you have an obligation to develop a strategy and prove return on the investment in time and money.

Return on investment is notoriously hard to measure because of the difficulty in determining what to measure and how to measure it. After you have determined that time and resources should be invested in order to increase influence, how do you find the time for this "extra" work? Is it really extra?

You can shape your role by studying the desired outcome and the values you place on influence. Depending on what you discover, you may find the role of enabler or curator yields better results than the role of content contributor.

Taking stock of current communities and conversations

Before you jump in and risk embarrassing yourself or your team, or worse yet, angering customers or internal departments, you need to analyze and understand the communities that already exist within your company and externally and learn how to interact or stand back.

As writers, we need to figure out whether we would be welcomed as members of the community or whether we should instead try to enable the conversation through our deliverables.

Realize that there are grassroots communities related to commercial products. For example, the website xmfan.com was started by regular people who are fans of XM Radio. Another website, IKEAfans.com, offers support help and an "IKEApedia" area that includes assembly manuals. When you approach a community started outside of the company, like these two examples, be cautious of whether you would be considered an intruder or outsider.

An excellent example of a writer getting productively involved with his community is Bob Bringhurst, the lead writer for Adobe InDesign and InCopy. He writes a blog at blogs.adobe.com/indesigndocs with announcements, buried treasures, and bonus documentation, to name a few categories. He is active on the user forums and uses his blog to talk to community members and link to others' blogs, such as this InDesign link round-up entry from 2009, "Best InDesign Links of May"[49].

You could follow different models for participation in communities. Here are three possible models for the technical writer's role:

- A "personality" who delivers content. In this model, the writer participates actively in the community.

- A "sage on stage" who instigates participation opportunities. In this model, the writer is like a stage hand who needs to understand the inner workings of the community theater to put on a good show. We can enable conversation through our user assistance without being the star of the show.

- A "bystander" who listens to the conversation. In this model, the writer listens to many communities and learns from the interactions of others, bringing that knowledge back into the technical documentation or passing it to the appropriate groups.

Anne Zelenka has written about the *information age* versus the *connectivity age* on the Web Worker Daily website[90]. I believe this distinction is an important one for the participation models that writers can choose. Are you an information worker or a connection worker, and does your corporate culture support you more in one model or another?

Anne describes the building of Microsoft as an example of the information age. The currency is always monetary, and the information that Microsoft created begat monolithic software. Google, on the other hand, has built a business model based on human behavior, making products that use attention, such as Adwords, to generate value.

If you can find a way to draw attention to your online help with conversation or community inclusion, you have moved your user assistance into the connected age. Be aware of areas where your company may already be moving towards the connected age.

If you have observed that most user communities do not publish comprehensive tomes, you are correct. As Andy Oram of O'Reilly Media says, people who respond to email lists do not even think they are writing documentation.[4] However, because a search can find the information and the information can answer someone's question later, it actually is documentation.

As Darren Barefoot said in his talk "Everyone's a technical writer" at DocTrain West in Spring 2008, the "people formerly known as the audience" are already documenting your products in ways you could never hope to cover yourself with a team of technical writers.

Instigator or enabler of conversation

An instigator provides a starting point for a conversation, perhaps by communicating a controversial decision or a highly debated strategic choice. A writer in an instigator role should know custom-

---

[4] http://www.oreillynet.com/onlamp/blog/2007/01/why_do_you_contribute_to_commu.html

ers' business needs and be well-connected with those he or she plans to talk to online.

An enabler understands the underlying concepts of a product or service well enough to help others understand those concepts as well. An enabler gives a community the authority to make decisions or provides patterns that help a community develop and grow.

Whether you're an instigator or enabler, you can repeatedly gather knowledge from communities and conversation, then bring it back and incorporate what you've learned into the documentation. Ideally you'll "give back" to the communities that provided the knowledge to you by continually improving the content offerings.

It may seem that an extrovert would be the instigator of conversations in documentation and an introvert would be the enabler of the conversations. Introversion and extroversion refer to how a person draws energy – either from an external source such as interaction with other people, or from an internal source, such as their own thoughts and imagination.

However, in the realm of social media, introverts often find that online social networking provides the right balance of encouragement to reach out, along with the ability to turn off the stream of constant communication and regroup for a period of time. Therefore, the role of an instigator does not need to be reserved for an extrovert. In fact, an extroverted instigator may not be a good blogger because good writing often requires introspection and reflection. But either personality type can take on either role and handle tasks within each role.

Building a strategy

As with all technical documentation projects, you should plan before implementing any changes to your documentation or approach.

However, when trying out social media tools, you need to be willing to spend time experimenting. Grassroots efforts that fly under management's radar for a while might give you the biggest payback.

In a blog post titled, "How Much Time Does It Take To Do Social Media?"[65], Beth Kanter, a social media consultant to non-profits, describes participation in social media as having five phases: listen, participate, generate buzz, share content, and create community.

A simplification for our approach involves four phases:

1. **Listen**
2. **Participate** in the conversations (and continue to listen)
3. **Share** content
4. **Lead** the way by offering a platform for conversation

A conversation enabler will work best in the first two phases. A conversation instigator can work in all the phases. I believe this phased approach yields better results than jumping straight to offering a platform.

You may notice that I skipped over Beth Kanter's third phase where she suggests that you should generate buzz. I don't think that generating buzz or media attention is necessarily the realm of technical documentation, though it might be if you're a lone writer bridging the gap in technical marketing documentation.

Listening phase

In their ebook, *Getting to First Base: A Social Media Marketing Playbook*[3], Darren Barefoot and Julie Szabo say that technical writers are excellent blog monitors and can easily handle multiple notifications from multiple news feeds or RSS subscriptions. If a technical writer also maintains a blog and online presence, commenting on others' blogs in the field is a great way to join the conversation, However, be aware that commenting on blogs moves into the participation phase.

Tools for listening are subscription tools such as RSS feed readers, Google Alerts, Twitter Search, Technorati, and Yahoo Pipes.

Another way to listen is to spend a day with technical support listening to their calls. This literal listening exercise can be eye opening if you've never spent a day on the customer line before.

If your product is sold on a website like Amazon, you can also find conversations in the reviews, review comments, and blog entries.

Participation phase

In this phase, you can blog and comment on blogs. The authors of the book *Groundswell*[19], Charlene Li and Josh Bernoff, suggest that you write five entries to see if you have the ability to stick with regular blogging. Internal blogging is also a good starting point, where the blog can only be read behind a corporate firewall. Other participation ideas include:

- Find a wiki and contribute. Start with small tasks such as editing or tagging.

- Offer a comment mechanism built into your help system.

- Offer a survey on your help system.

- Start tagging sites relevant to your product on delicious.com, reddit.com, or stumbleupon.com.

- Offer a live chat or take a turn staffing the live chat if you have the knowledge.

- Join or moderate a customer forum or board that has private messaging where one-on-one conversations can occur.

Content sharing phase

The content sharing phase may be as simple to implement as adding a Creative Commons license to your content to allow for share and share alike. Licensing your content with a Creative Commons license declares that your content is reusable and describes the conditions and circumstances for that reuse.

There are different licenses available for you (or your legal department) to review. Instead of reserving all rights to your documentation, you reserve some rights, but also free the content to be used by others. For example, the Adobe help content offered with the Technical Communication Suite has a Creative Commons license, Attribution Non-Commercial Share Alike, as indicated by this icon:

On the Creative Commons[8] website, you can answer a few questions to help you determine the right license for your needs. Questions include, do you want to allow commercial use of your work, do you want to allow modification to your work, how can others share your work, and what is the jurisdiction of your license?

Another way to share your content is to offer an RSS feed that others can embed in their websites. Web services like feedity.com can help you create an RSS feed from any web page.

You can share content by uploading photos or screenshots to a photo sharing site such as Flickr, or you can upload videos or screencasts to YouTube or Vimeo. Another great example of sharing is the Autodesk Civil Engineering Community.[10] Users can join the community and download templates, styles or shapes.

---

[8] http://creativecommons.org
[10] http://civilcommunity.autodesk.com/content/

I have found Creative Commons licensing extremely helpful from another perspective, in my work as the OpenStack technical writer and documentation coordinator. I found that bloggers were writing about OpenStack and had licensed the content with Creative Commons. I reached out to the bloggers, invited them to incorporate their blog entries into the OpenStack manuals, and created several good authoring relationships.

My role felt like an acquisition editor, where the blog entries about OpenStack were the submissions, and I could cull through them for real gems. In fact, one blogger whom I contacted, Ken Pepple, later became an O'Reilly book author based on his blogging and sharing with the OpenStack docs site.

Platform or stage phase

If the first three phases yielded connections and a sense of community that you could tap for growing community-based documentation, your final phase might be to create a wiki for your product, either as a supplement to your documentation or as the new home of your entire user assistance content.

Alternatively, if users and customers have enough to say about your product that you think it would be helpful to the community to host a blog, the result of the first two phases might be to offer a blogging platform.

You might find that a forum becomes a platform for developers to collaborate on projects. The EMC Community[11] offers blogs, forums, and social media opportunities on platforms such as Facebook and Flickr in one community site.

Another example of a social media platform comes from Sun Microsystems.[12] Two writers, Gail Chappell and Cindy Church,

---

[11] http://emc.com/community
[12] http://sun.com/communities

presented their experience with Sun at the 2008 STC Summit. Both their product and their audience were a good match for providing a platform for documentation through a wiki. Their audience was college-age developers in school plus industry-savvy developers in the workforce, all busily coding enterprise applications using Net-Beans Ruby, a free, open-source software tool for creating Ruby language programs and part of the NetBeans integrated development environment.

For their case study, contributions from the community were welcomed, both for the product and the documentation. The wiki did not take off as a platform for community-generated content as you might think, though. They used the wiki for providing drafts of documentation and delivering info quickly. Content on the wiki did not need to be perfect or in a strict template. Once a draft was posted to the wiki, writers then posted a message to the user forum to let them know they could review it. Once the content was reviewed and vetted, writers transferred it to the main website instead of the wiki.

One lesson to learn from their experience is that even though the writers provided a platform for community content, the model that was most valuable was using the wiki for drafting content then transferring the content to a non-platform-based deliverable.

Putting it all together

During any phase you can consider the level of your content itself. Any one of these levels may fit your business needs, but you can use Table 2.1 as a roadmap towards using social documentation to further more and more business needs.

Table 2.1 – Levels of Content

Level	Business needs
Level 1: Offer content on the web and ensure that search engines can find it.	■ Customers need to find information fast. ■ Customer support wants fewer expensive support calls. ■ Your department needs to raise visibility to prove value.
Level 2: Offer a subscription, such as RSS, to your content, tag your content, or share it in a new way such as in a screencast or on a photo-sharing, book-mark-sharing, or video-sharing website.	■ Customers want to know when something has been updated. ■ Immediacy of knowledge, such as "I want to be the first to know about new features for this product because I'm my company's expert in it."
Level 3: Enable comment, feedback, or a forum.	■ Consultants or professional services employees want to share what they have learned. ■ People want to meet each other in person, for example at a user group, to find answers. ■ Customer have scenarios to share with others in similar situations, perhaps using a tool in a new way. ■ Customers want a single point of contact and one voice.

Level	Business needs
Level 4: Include outsider-generated content.	■ Consultants or professional services employees want to tell you and others what they have learned. ■ Customers have scenarios to share with others in similar situations.
Level 5: Provide complete transparency at all levels and allow anyone to contribute.	■ Customers, internal or external, want a way to share unsanctioned knowledge for discussion and debate.

The documentation environment

What is the lifecycle of the products that you document? If your company releases products often, your users may expect documentation updates every release. A wiki may be the best way to fulfill this expectation. If the product releases only once every other year, you can invest in conversations during those longer release cycles by using a different format, such as customer forums, blogs, or podcasts.

What are your current technical publication deliverables? If your team still relies heavily on printed documentation, you might not have an efficient method for getting feedback from or conversing with customers. If you continue to deliver printed documentation, you will want to find online methods for enabling conversation and community. If you do not already have a method for gathering feedback on printed documentation, investigate the costs and returns of sending printed-format surveys or questionnaires.

Using surveys to analyze audiences

Surveys are yet another conversation point with your audience, especially if you have an integrated survey system tied into your help system. Collecting and analyzing survey data is a proven method for discovering the communities your readers already use to find answers. You can survey your audience in several ways, even inexpensively using services like Survey Monkey.[13]

Google Forms are easy to design, and when someone fills out the form, the data goes into a Google Spreadsheet as it is collected so you can see partial results before the survey closes. If you don't want to expend resources sending out your own survey, consider adding a question or two to surveys that are already sent to your customers.

Determining if your customers want a conversation

Look at the specific problems that your technical publications group is trying to solve or has had feedback about, and analyze whether applying social networking or notification technology would help solve those problems. Customers cannot always articulate their needs in specific requests for a certain deliverable like a wiki, and terms like "RSS feeds" and "wikis" may not be familiar to them. But if you asked them, "Would you like to know when new items are added to the help system?" they would probably respond positively. Also consider the prospect who is not yet a customer. What conversation or community techniques can build relationships that help with converting that prospect into a customer?

Analyze your audience to determine their pain points. Ask questions or conduct a survey, and pay attention to the demographic information. While gender and age are not always good indicators of an individual's behavior online, research by Bill Heil and Mikolaj Piskorski, "Twitter – New Research: Men Follow Men and Nobody

---

[13] http://surveymonkey.com

Tweets"[59], indicates that men use certain social media tools like Twitter differently than women do. Different age groups might have different answers for your survey, or not.

A survey run by two technical writers at Sun had an audience of programmers, most of whom were male. The writers wondered if age differences would result in a different "wish list" for video or screencasts. They found that no matter the age, all of the programmers wanted pretty much the same thing – examples and a web celebrity of sorts to listen to and learn from. Using the information they gained from the survey, the writers provided a business case for shifting their priorities from traditional online help to videos and screencasts.

If you have the resources, you might consider commissioning a survey of your customers using a tool like the Forrester Research Social Technographics tool,[14] which is described in the book *Groundswell*[19]. Figure 2.1 shows sample output from this tool.

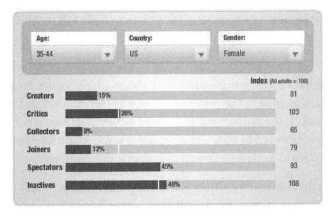

Figure 2.1. Data from Forrester Research Technographics surveys

---

[14] http://www.forrester.com/Groundswell/profile_tool.html

Creators are at the top of the ladder because they create content such as blog entries or videos. Critics are most likely to respond to others on email lists or forums. At the bottom of the ladder are Inactives, those who do not create or consume social content. If your study shows that many of your customers are Inactives, you may not need social media integration to serve your customer's information needs.

Determining your role

You can ask several questions to help discover if you are an enabler of conversation, an instigator, or both. Depending on the number of users of the products that you document, there may be a place for the technical writer in the community. For wiki documentation especially, it is important that you become a conversation starter, unless there's a large community base already. Several factors influence this determination:

- Is the community small, and if so, can you easily become like the community members? Or is there a conflict of interest that might prevent you from direct communications?

- What motivates the community to use your product? If community members are passionate about your product because they selected it, the conversation may be easy to start and continue. If the product was purchased by upper management and implemented by community members, you might be more effective as a conversation enabler than a starter/joiner.

- Is the community getting paid to implement the product? If so, will they view your presence as competition? In this type of situation, it might be better to enable conversation.

- Has another department in your company started a conversation? If other departments internally are in fact responsible for responding, be careful about how your involvement might be interpreted. Talk to others internally before inserting yourself as a community member or you may appear to be strong-arming others. A slow, easy beginning is best, and practice and hands-on experience are your most valuable assets.

Whether you're enabling a conversation or are part of the conversation, be professional, but also personable and genuine.

Also, expect to face the inevitable negativity or conflict that can occur within a community. You should face it, communicate openly, overcome negative discussions, and address negative situations.

Corporate policies for social sites

Read your corporate blogging policy or public discourse policy or social networking policy. If your company does not have such policies, research examples from other companies and consider whether they fit your culture and how you might adjust them for your needs.

IBM has a social media policy[15] that is considered to be the "gold standard" for these types of employee guidelines. For non-profit organizations, Appendix B contains the policy for online discussion from Easter Seals which encompasses many different types of online publishing areas.[16]

These policies have many commonalities for trustworthy, valuable online representation. They give you advice like: be genuine, always identify yourself and the company you represent, be professional, talk about what you know, cite sources and acknowledge others' contributions, obey copyright laws, offer value, don't pick fights, and respect your audience. And they ask the employee to abide by the policy and use good judgement in all interactions.

---

[15] http://www.ibm.com/blogs/zz/en/guidelines.html

[16] http://beth.typepad.com/beths_blog/2008/04/nonprofit-blogg.html

Additionally, study your corporate culture to determine how much openness and transparency already exist. Look for areas where transparency and openness are rewarded and not feared. Find ways to mitigate the risks that you may take by being transparent in your communications.

Do not contribute if your contribution could be interpreted as interrupting "real" work (for example, you may want to avoid posting long responses during work hours).

Analyzing user expectations

User expectations and corporate culture play an important role in evaluating whether your products and documentation set could benefit from social networking and conversation. Be aware that if your products contain open source software, customers will probably expect open documentation that they can edit or remix themselves.

If products are not open source, but users are accustomed to receiving early revisions of the documentation, those customers will also have expectations for how the content will be handled. A writer in an enabler role would do well to keep up with those expectations.

Defining business objectives

The book *Groundswell*[19] discusses a four-step approach to getting corporations into conversations: People, Objectives, Strategy, and Technology (POST). I believe this approach can apply for technical publications and social media as well. The business objectives that your department typically aligns with are your objectives for any social web exploration as well. I especially appreciate that technology is the last step in their approach.

Before you attempt direct conversation with customers or try to enable more customer-to-customer interaction, ask yourself what you hope to achieve. Articulate the business goals so that you can communicate them to your manager and others. Also, examine how realistic your goals are given the time and resources that you have available. Keep in mind that you may have to use your own time and resources to prove the value of your proposal before you can gain the confidence of those who measure business value and make decisions. Many of the Groundswell case studies involved a couple of guys and an under-the-radar server installation.

To discover where you might gain business perspectives in different areas, such as Operations or Finance, look at the metrics in the section titled "Recruiting others" (p. 188).

Leave room in your social media business strategies for unintended consequences. For example, one group of users may not want "how to" information coming from a social deliverable; instead they might want reference information that they can add to. Another group may not be interested in status updates that answer "What are you doing," but instead want to know "Where are you?"

Agile development

What development method does your team use? Agile development refers to a group of software development methods that incorporate user involvement into development. On an Agile team you use time-boxed iterations to deliver small features of potentially shippable software at the end of each iteration.

Central to many Agile shops are Scrum project management and self-organizing teams with cross-functional team members, including technical writers embedded on development teams. The way software is developed usually influences the way that the technical publications team works, and Agile methods may lead to bigger changes than most.

If you work in an Agile development environment, however, your team has probably already had to adjust the publication release cycles, information development methods, and the amount of documentation that is promised with each release, depending on the staffing challenges that you face because of Agile adaptations.

Sarah Maddox has a wonderful two-part blog post, "The agile technical writer"[67], in which she discusses her typical day as a technical writer working on an Agile development team. Much of her work involves IM conversations and editing in a wiki environment, using the wiki product she documents as the documentation deliverable.

Mike Wethington, a technical publications manager, says that the days of introverted technical writers expecting to be handed specifications and isolate themselves while they write the documentation are gone. Agile methods require internal team communication that resembles a conversation, with daily team standup meetings and collaboration with team members on writing tasks. The tools of the social web are critical to the speed, efficiency, and focus that a disciplined Agile team must have in order to succeed.

Identifying key team members

For those who are interested in being a representative voice for your department, for the technical documentation, or for the product itself, determine the methods that team members can use to begin conversations. Each team member should have a consistent, professional, and approachable online persona. Search for your first and last name on Google and other search engines and try to envision what you represent to customers, both internal and external to your company.

Examine existing community starting points for your customer base. Begin the conversations in a non-intimidating manner, such as commenting on blog posts. Always represent yourself honestly, draft a standard disclaimer, and determine when you will apply the disclaimer. For example, you should always identify yourself as a corporate representative when the blog post addresses a topic specific to your company.

For some team members, nothing could be more terrifying than losing anonymity while writing technical documentation. Because identifying oneself online can have negative consequences, do not require this type of involvement as a term of employment for existing employees. Writers may want to maintain two personas online, a personal one and a professional one. Behind-the-scenes and connector roles do not require the loss of anonymity or privacy. Rather than being directly involved in the conversations, writers in these roles facilitate conversations and community by incorporating them into the documentation system.

Piloting internal solutions first

It is very easy to stumble while practicing the art of conversation and community so you might want to try internal conversations first. For example, start an internal wiki so that you and your team can become accustomed to the new way of working and communicating.

Stewart Mader, author of *Wikipatterns*[21], touts the value and necessity of using a pilot project to start wikis, and I believe this approach works best for implementing any type of social media. This technology is still relatively experimental, and you should allow your team to experiment to find the strengths and weaknesses of any new delivery mechanism or community system.

3

Community and Documentation

The scale of the global community that is going to be able to participate in all sorts of discovery and innovation is something that the world has simply never seen before.

—Thomas Friedman, *The World is Flat*[10]

This chapter explores the idea that a small group of people who have a sense of belonging in an online community can provide content much like a technical writer does. Regardless of their background, education, or training, more people are becoming providers of technical information on the web. The gift economy, where altruism motivates contributions more than monetary gain, is partially responsible for this shift. Chris Anderson, author of *The Long Tail*[2], also wrote the article "Free! Why $0.00 Is the Future of Business" for Wired magazine[41] in February of 2008. In the article he describes a taxonomy of "free" that contains many models including the gift economy.

In *Naked Conversations*[30], bloggers Robert Scoble and Shel Israel state "Altruism turns people on even more than making money." Combine these altruistic motivations with a publishing system that has a low barrier to entry and you get a whole new way of working and writing.

If anyone can be a writer, what communities of writers have formed, and what have they accomplished? Why would you leverage someone's sense of community belonging for documentation instead of hiring professional writers? How do you form communities online and in the real world?

What is a community?

Community has many definitions, but an *online community* has specific elements that shape its definition. From a blog entry by Forrester Research analyst Jeremiah Owyang comes this definition:

> An online community is: Where a group of people with similar goals or interests connect and exchange information using web tools.
> —*Defining the term: "Online Community"*[74]

As you might expect from a community researcher, Owyang says that his Twitter followers helped him shape that definition.

Howard Rheingold defines a virtual community as:

> ... when people carry on public discussions long enough, with sufficient human feeling, to form webs of personal relationships.
> —*The Virtual Community*[28]

Think of the online communities you belong to. You know there are common elements of discussion, agreement, core values, and unifying goals that help you identify it as a community. Relationships are the makeup of communities. Connections with each other establish relationships. Not all relationships in a community have to be positive – even contentious debate can help a community meet its goals as long as the relationships remain respectful and intact after discussions.

Although in-person collaboration is an aspect of communities and documentation, much communication happens only online until community members meet at a conference or a networking event. Much of the communication that helps you feel like you know a person in the community happens online long before a face-to-face meeting. It's that in-person or online conversation and public discourse that forges the relationships that create a community.

What's not a community?

Sometimes the tools that help create the connections are mistaken for the community itself. Community is built from the connections, not the tools. For example, a blog with comments might constitute a conversation, but it does not constitute the community itself. This distinction is important for a few reasons. First, it can help you choose which tools you can use to form a community. Second, it can help you understand why you might not have a community on your hands, even

> Community is built from the connections, not the tools.

though the right tools are in place. And it can help you understand why a wiki has few edits, yet many hits. Perhaps the sense of belonging to a community with a specific purpose and set of goals and the motivations that follow is not yet present. Or perhaps the technology layers are not familiar, so people don't feel comfortable in an online community until they make sense of the tools for communication and interaction.

Motivations for writers and online communities

Why are any of us interested in documenting a complex product or process? It's possible that the core of our motivation is recognition or reward, measured in terms of money or success. Or maybe we think that writing and communicating with text, images, audio, or video is a great way to make a living. But an underlying motivator for many technical writers is the desire to help others learn. By teaching, you learn twice. By joining a community, being a community member, and looking for places where you can either contribute or motivate others to contribute, you are empowering collaborative efforts unlike any seen in the past. What I am observing more is that community members who do not have a writing role still want to write or communicate information about the product or service.

I can identify with the desire to help people through my work with the One Laptop per Child project. My kids and the kids in my son's preschool classroom inspired me to work for the first time on an open source project. My first experience with Sugar, the open source operating system for the laptop, was an emulation on my laptop, and I was floored by its elegance and simplicity, created by a community of developers and usability experts giving their time to the project. I was inspired by the high quality of their offering and wanted the documentation to match it.

I have found my volunteer work to be invaluable as a learning experience and exercise in connecting to others. But I will also admit that one of my motivations has been to experiment with publishing and book promotion, with an idea in the back of my head that FLOSS Manuals could hit upon a huge best-selling book idea.

Before the OLPC book sprint in August 2008, the FLOSS Manuals community had quite a nice discussion about money and free documentation. Income from book sales is typically used to further fund FLOSS Manuals' goals as a non-profit community – invest your gains to further your aims.

So while writers are accustomed to earning money directly for writing, they might find that they are motivated not by the money but by the additional activities the money can fund. For example, profits from a book sprint can be used to fund future book sprints.

New roles for writers

One of the roles I learned about while writing for a wiki is "community manager." In many wiki environments, the writer can be the "content curator," someone who assembles collections based on themes. But if your goal is to build and grow a wiki, you need to build and grow the community for it first. People who feel they are part of a community may share knowledge as the community becomes a community of practice.

Clay Shirky, in a talk[1] about his book, *Here Comes Everybody*[31], explains how a community of practice can form quickly when there are online tools that speed up teaching specific techniques and learning from others. A community of practice uses people as conduits to transfer knowledge. As those who study knowledge manage-

---

[1] http://cyber.law.harvard.edu/interactive/events/2008/02/shirky

ment have learned, workers spend a third of their time searching for information, and they are five times more likely to turn to experienced co-workers for information rather than read a book, thereby shortening their learning curve and hopefully avoiding mistakes.

You have probably heard of the phrase "train the trainer." You may even imagine that your role as a writer is to write to the person who reads enough to learn a product inside and out and become the in-house expert for it. As an extension of that writer's role of training the trainer, a writer may become a leader within the online community, teaching other community members. This trainer role may be highly motivating to some.

Particular types of users are more likely to seek out online communities for information – competent performers, proficient performers, and expert performers, as defined by Dreyfus and Dreyfus in *Mind over Machine*[8]. In their "Model of Skill Acquisition," these types of users are separated from the novices and advanced beginners, who represent the first two stages in the model. Advanced users are more likely to demand useful examples, instructions for getting results or achieving a specific goal, and advanced troubleshooting information.

When managing a community of experts, realize that the rules of the community can make entry difficult. Consider that some communities have an unspoken "no whining" rule, which creates a very different atmosphere than a community that has a "there are no dumb questions" philosophy.

"Free as in freedom" documentation

Why do people contribute to documentation for no payment or even unknowingly since they don't consider themselves to be writers? Andy Oram, senior editor at O'Reilly Media, studied contributions to documentation by surveying people who contribute to online forums, email lists, or wikis. His report, "Why Do People Write Free

Documentation? Results of a Survey"[73], contains many findings, some of which are surprising:

- People generally don't think that they were contributing to documentation.

- People don't think of themselves as writers, despite the fact that the communication is written.

Because the people surveyed were not paid directly for their written contributions, their contributions could be considered free as in "no-cost." In open source circles, "free" can mean "freedom" rather than "no cost." "Free" documentation, however, represents a shift away from these two definitions.

The email/Internet/search engine infrastructure that is in place enables all of us to feel like we are getting and giving information for free. In reality, a small group of people who have computers and pay $40 per month for high speed Internet access offer their time and knowledge in information exchanges in the hope that they will be repaid by others' time and knowledge, through recognition, by achieving a sense of belonging, or by gaining back time through efficiency gains. These motivations for contribution (reciprocity, reputation, attachment, and efficiency) in any online community are documented by Peter Kollock in his paper, "The Economies of Online Collaboration: Gifts and Public Goods in Cyberspace"[66].

Free as in beer vs. free software

In the open source world, there is a distinction between "free," as in no cost, and "free software," which means "users have the freedom to run, copy, distribute, study, change and improve the software." The term "open source" is frequently used as a synonym for "free software," though purists claim a distinction in intent. For more, see "What is free software?"[52]

For some, payment is in yet another free, no-cost form. Perhaps you could even go so far as to say that their payment is in happiness. Tara Hunt, author of *The Whuffie Factor*[14] and blogger at http://-tarahunt.com, discusses the happiness factor in a 197-slide deck (offered for free!).[2] The following four items represent what Tara calls "pillars of happiness":

- **Autonomy:** People want to personalize their experience, have choices, and have an open and transparent environment.

- **Competence:** People like to feel they're good at what they do.

- **Relatedness:** People want to connect with others in similar situations.

- **Self-esteem:** People like to feel confident in their knowledge and relationships.

Joining a community

This new collaborative economy will affect technical writers who are paid for documentation. Part of the new role for writers in Web 2.0 is becoming a community member, being a contributor, and being a reader of contributions from other community members.

Community membership is a hands-on activity that you do best when you have practice at it. It is not easy to get experience, but it is necessary. You must read the online arguments (flame wars) and observe other's reactions to know how to handle them better each time they happen. You can observe leadership passing hands and the delegation that occurs. But observation does not always teach you how to step into a leadership role or delegate well.

You can learn the ebb and flow of a conversation and become a meteorologist for the "social weather" that's ongoing in a community. For an example of social weather, do what Clay Shirky

---

[2] http://www.slideshare.net/missrogue/happiness-as-your-business-model-414463

writes in his description of the course with the same name at New York University. Simply make some observations next time you walk into a restaurant. Is it noisy or quiet? Slow or busy? Are there couples or groups dining? In a restaurant you have visual and auditory cues to give your inner meteorologist a chance to assess the social weather. In an online community you need to understand the cues that occur in writing, in emoticons, and in frequency and intensity of updates to content.

At the Internet Research 3.0 conference in October 2002, Alex Halavais described a deep dive analysis of bloggers' discourse:

 By measuring changes in word frequency within a large set of popular blogs over a period of four weeks, and comparing these changes to those in the "traditional" media represented on the web, we are able to come to a better understanding of the nature of the content found on these sites. This view is further refined by clustering those blogs that carry similar content. While those who blog may not be very representative of the public at large, charting discourse in this way presents an interesting new window on public opinion.
— "Blogs and the 'social weather'" [56]

While this concept may sound new and exciting, it is quite 20th century. In 1912, A. A. Tenney analyzed newspaper content to determine public opinion ("The scientific analysis of the press" [86]).

Tara Hunt's pillars of happiness are a good set of guidelines for testing social deliverables. When community members can voice an opinion, give an honest review, and build an article, diagram, picture, or video, they feel autonomous and happiness follows.

Kathy Sierra calls creating competent users "helping users kick ass," and she has written valuable blog entries about how to do that.[3]

In the area of relatedness, a writer doesn't necessarily have to be a community member to enable people to talk to each other and meet in person more easily. Organizing tweetups (in-person gatherings for Twitter users) or user group meetings is an extension of the community manager role, enabling people to feel related to one another. Documentation that enables relatedness, such as helping wiki writers talk to each other via comments or "talk" pages, contributes to that pillar of happiness.

I found that after moving OLPC content to the FLOSS Manuals wiki, I still communicated project information via the OLPC wiki.[4] When I organized the OLPC book sprint, I created a request page on the wiki to invite community members to the event. Volunteers then contacted me through my wiki "Talk" page, which sent me an email notification while still keeping our discussion open for any project contributor to see.

Growing a community

Create guidelines and a central purpose for the community if you are going to grow one organically. For example, some online communities subscribe to one overarching rule: No whining. Others may set a basic rule like: The only dumb question is the one unasked. These two communities will attract different members.

Also, after you identify an established community you want to join, realize that it may take a direction that you hadn't considered by changing its goals or changing some of its rules or culture.

---

[3] http://headrush.typepad.com/

[4] http://wiki.laptop.org

Rather than trying to single-handedly grow the community, delegate to and recruit from the established community members for tasks that are not in your area.

Communities may have patterns and anti-patterns much like the wiki patterns that Stewart Mader identified in *Wikipatterns*[21]. In online communities, people can block progress on community projects such as coding open source software. For example, someone might offer to address a bug but then doesn't follow through. The lack of follow-through becomes a block for anyone else who wants to make progress on the bug. Lack of documentation can also hamper community efforts. Ever-changing or ambiguous licensing of code may cause a community to abandon it if the license is not easy to use or not in line with the community's vision.

Realize that benevolent dictatorship may not be the correct approach for a community. A benevolent dictatorship usually involves one leader who maintains a stance of always doing what's best for the community, but maintains control over all decisions. Keep a balance as you foster community interaction, especially if you become a leader. Realize that everyone can have a role or several roles at once, both leading and following.

Software management books author Gerald Weinberg once said, "No matter what the problem is, it's always a people problem." Remember that technology layers cannot fix basic trust issues or bad behavior. Rather than adding administration layers, complex access control lists, or even check-ins, talk through the people problems. And if you are the benevolent dictator, be ready to allow the community to self-govern if that is the right direction for it.

Real-world events

One reason people join an online community is the in-person interaction that can happen at professional conferences or networking events. They may join a community related to their profession or to the tools needed for their job, or a community related to a fun recreational interest. It follows that people want to meet others in person at meetings and gatherings. These events not only enable you to communicate with your company's customers; events help the customers meet and talk to each other.

User groups and focus groups are examples of real-world meetings that have a single goal in mind. Unconferences, barcamps, and meetups are another type of real-world event where people with similar interests and goals meet to share information. Book sprints borrow from these camp concepts to create a one-week event of collaborative authoring that generates a book.

Unconferences, barcamps, and meetups

An unconference or barcamp is usually defined by its lack of definition. The morning of the event, attendees gather at one place in front of a white board or poster and offer to share their knowledge in presentations or demonstrations. The unconference has no agenda until session ideas are placed on a white board, often using Post-it notes, in certain time slots so that participants can find sessions to attend. Sometimes a wiki page can help shape the agenda before the event, but the board at the event is the "final" schedule.

A meetup is a prearranged group meeting. The online social networking site meetup.com is one service that facilitates organizing those meetings. A combination of meetup and Twitter, a Tweetup is an in-person meeting organized using Twitter for publicity and invites. Sometimes these events are intended for casual entertaining, other times they are organized for professional networking. The main point of these types of gatherings is to take advantage of the higher fidelity conversations that in-person events offer.

How can you plan for such a purposely impromptu event as an unconference? First you have to realize that the attendees are going to make or break the event. I had little experience with planning an unconference when I was asked to plan one that would take place during the 2008 DocTrain West conference. With the encouragement and help of one person who was certainly going to attend, I was able to grow the invitation list with like-minded individuals. We ended up with some of the most interesting attendees, and one person said it was the most useful session of the conference to him, a compliment considering that the unconference was embedded within a highly valuable conference with a devoted group of attendees.

Learning from Mozilla: Interview

Janet Swisher and Eric Shepperd work at Mozilla Corporation on a large wiki at developer.mozilla.org.

1. **How are you aligned in the company? Support, marketing, training? What is your job title?**

    The Developer Documentation team is part of the Developer Engagement group, which is part of the Engagement team, which is, basically, marketing.

2. **What types of content do you deliver?**

    All of our content is on the web, primarily technical documentation (both references and tutorials), but we are increasingly trying to do demos and other types of training and developer education materials (videos, for example).

3. **How is your content licensed and how was the license selected?**

    Our content is licensed under CC:SA. Historical code samples (from prior to August 20, 2010) are under the MIT license, while all newer code samples are public domain.

    There are occasional articles that are under other licenses, depending on the source of the content, but those are far and away an exception to the rule.

4. **How many collaborators do you work with regularly?**

We have two on-staff writers (with another hopefully about to be hired), and a half-dozen or so regular, substantial contributors. We then have another dozen or so contributors we see fairly often, plus the next tier of people that contribute only occasionally.

There are probably 10-20 contributors who are trusted enough that their content doesn't get substantially reviewed other than to ensure technical accuracy, since they're trusted to get the style and the like right.

We also have one or two non-employee contributors that have privileges above the norm, including one in Germany that is the only person outside Mozilla Corp. to have delete privileges in our wiki.

5. **Do you hold sprints and if so, how many people are active in sprints?**

We do hold sprints, alternating between in-person and online, virtual sprints. Janet can speak to this more than I can.

Janet responds, "For in-person sprints, it has ranged from 5 to 10 people. For online sprints, around 20 is typical for the number of people involved at some point during the sprint. Online sprints are 2 days, over a Friday and a Saturday. In-person sprints are 2-3 days, with the timing depending on circumstances."

"Most of our contributors are in Europe or North America, though we have one very active contributor in Australia, and have recently had a couple in Taiwan."

6.   How many page views do your pages get?

We average around 600,000 to 800,000 page views a day (depending on the tool you use to count them).

7.   What is your tool chain and how is it created? Internally, with outsourcing, through the community?

Our content is almost all hand-written directly in the wiki, with probably about 75% of the writing done by myself and Janet, and the rest done by community members. The community generally does more with fixing up existing material, correcting our mistakes, and so forth than creating original content, with the exception of a handful of people that contribute in more detail. The wiki is currently MindTouch, but Mozilla is in the process of developing an in-house replacement, Kuma.

We do have one utility that generates initial outlines of references for certain types of content, but that's the only automation we have, to speak of. This tool was developed by a community volunteer. Another volunteer created a tool that helps us keep additional metadata about items in the change-tracking system (Bugzilla) that affect developer docs.

8.   Anything else you want to share about your particular situation?

I think I included it all in my rambling above. :)

Book sprints

A "book sprint" is collaborative authoring in a short time, such as 3–4 days, with the express goal of publishing a book. That sums up the goal of the book sprint as devised by Tomas Krag, whose interests lie in the use of technology for developing nations, and who is much more interested in the distribution of content and ideas than protecting his rights as an author.

Adam Hyde, founder of FLOSS Manuals, was keen to explore the idea of an extreme book sprint, which compresses the authoring and production of a print-ready book into a week-long sprint. The first year after the birth of the book sprint concept, Adam and FLOSS Manuals experimented with several models of the sprint.

An entire book about book sprints is available on the FLOSS Manuals website. It offers guidance for running your own book sprint. This section reuses some of that content to spread the book sprint ideas and concepts as a collaboration technique.

Book sprint planning

Although the goal of the book sprint is to produce a book in one week, the planning of the event often determines its success. This section details some of the planning you should do in advance of the book sprint dates. Although a FLOSS Manuals book sprint uses the FLOSS Manuals wiki tool, you can use this advice with any collaborative authoring tool.

Scope

What topics will the book cover?

When determining the scope of a book, you might experience a "chicken and egg" scenario. You might want to plan the scope and then invite people to attend, but at the same time you may wonder whether it is better to invite people to attend and then let them decide the scope. Adam Hyde thinks that you can strike a balance. You can decide on the overall topic of a book, such as an existing software product, before inviting anyone. The writers participating in the sprint can then help brainstorm the details of the content. It's easier to motivate people if they have input early in the process.

Target audience

Along with the scope comes the target audience, and one decision certainly shapes the other.

Who do you want to read the book? What are the job titles they hold? What organizations or groups do they belong to? Are they students in a college or industry professionals in the workforce? What other publications might they read, or what subject matter should they already be familiar with? What is their general level of computer literacy? What operating systems do they use?

Answering these questions should also help you identify potential participants in the book sprint as writers, editors, or technicians who can serve as subject matter experts.

Ideally, you can include a person from the target audience in the sprint as a document tester. For example, if the target audience is "newbies," then invite someone who is new to the product. Instead of trying to imagine the level at which to write the material, you will find more value in having a member of your target audience in the room to look the experts in the eye and say "I don't understand what that means." The experts may then have to recalibrate their tone. This is not to say that the target audience member is always right, but the experts have to justify their position when challenged, which leads to better content.

Invitations

An important first decision when planning an event is whether you will select and invite a limited number of attendees or let the attendee list grow by allowing those initially invited to invite others. After you determine your core set of participants, choose a date and location that should work well for most of them. After you set a date and location for the core set of participants, you can open the event up to others.

With the FLOSS Manuals OLPC and SugarLabs book sprint, we had a last-minute request to attend the book sprint from a journalist located in the UK. Although she didn't have firsthand experience with the XO laptop, she was interested in the One Laptop per Child project and wanted to attend the book sprint in order to write a freelance journalist article about the experience. Although we were intrigued and flattered, we declined her request to attend because we wanted writers who could contribute immediately and also wanted to avoid distractions for the other attendees and participants. It turned out to be the right decision for all involved since we were a very focused group.

Draw up the outline

You perform two types of planning for a book sprint: planning the content and planning the event itself.

A topic-oriented, single-sourcing approach to content means that each chapter is a standalone piece of content that can be reused in many different remixes. This approach also means that the content can be used in a printed book, a PDF file, a collection of browse-able HTML files, or embedded in a website as if it were directly created in that website. Planning with reuse in mind is important to the success of the sprint.

You should choose reusable content that can be completed in the allotted time for the book sprint. Part of the time spent in a book sprint is reusing or rewriting what else has been written on the topic. Identifying content that is already written ahead of time is part of your planning process. What books, websites, or wikis already discuss the technology on which book sprint focuses? Can you reuse any of that content based on their licensing of the content?

After you have a proposed audience and scope for a book sprint, take the time to talk about your ideas with others who are interested in reading the content. Make sure you have buy-in for the audience and scope for which you are planning, and revise as necessary until you feel ready to start outlining the book. In a few book sprints, the outline was completed the Sunday just before the week of the sprint.

Reuse existing content

After you have created the outline, you should spend time finding as much existing content as possible. Reuse, reuse, reuse. Search the web and book stores for related content and write to each of the authors asking if you can reuse their content.

If material is available, and it is under a liberal license – the one you want to use – ethical and fair treatment of the content means writing to the authors, not to ask their permission but to ask for their "blessing" or endorsement to reuse the content. Many authors like to know where their content is reused, and you might find that you have a new, enthusiastic contributor coming on board as the result of your communication. At the very least they will probably tell other people about your project, and that provides good outreach for you.

Getting existing material is important not just because it can contribute to the total content of the book, but because it is motivating for the writers to see that some of the work has already been done before they start. The writers might throw most of it out, but by then, the content has already had its motivating effect.

Book sprint logistics

You need to plan the dates, location, and participants in the sprint before you get very far. Usually, the location and schedule of the participants will determine where and when the event takes place. Location and date selection depend on each other because some locations are not going to be available for certain dates. Travel time, distance, and cost, as well as visa considerations, are also part of the equation. If participants need a visa to come to the U.S., at least a month lead time is necessary.

Transportation

How will everyone get there? Will you find the funds for travel or will the participants need to cover some or all of their travel expenses themselves? Travel costs are often the largest expense. The following sections provide an example of the travel expenses for a book sprint with about ten to twelve participants in Austin, Texas.

Airfare:

- Airfare from Amsterdam $1,000
- Airfare (for two people) from Boston $800
- Airfare from Wisconsin $400
- Total airfare $2,200

Six additional participants were all within driving distance of Austin. We did not have to pay for parking at the host location, and we did not reimburse mileage for driving, which would change the total travel costs.

Car rental:

- Two cars, $100 per for a week, $200 total

We did not rent cars for this book sprint. Driving people around once they were in Austin was my duty as a host and logistics coordinator, but we could have rented one or two cars (at about $100 each) for the week and made our lives much easier. Good logistics planning allows writing to be the focus of the week.

Location and other venues

The venue needs to have a single table where everyone can work. The sprinters need ample power sockets (or lots of extension cables), and you need a good Internet connection, preferably wireless. All-hours access to the room is helpful. Some host locations may want participants to sign a release so the host is not liable for damage or injury. Signing a release should be fine – nearly all of the planning is based on good faith in others anyway, so suing when the location is donated would be poor form.

Location costs were $0.

Accommodations

Where will everyone sleep? Will they pay for their own accommodation or will you find funds for it? Is the accommodation sufficiently close to the book sprint venue so that participants don't waste energy and time traveling every day? Do heavy snorers, smokers, or light sleepers need their own room? Do you need to have a space to prepare and cook meals?

The hotel rate was a corporate rate, and the total cost was $1,333 for two rooms for five nights, with the rooms shared by four participants. The hotel was within walking distance of the sprint location.

Food

What is everyone going to eat and who will pay for it? Are there any special food requirements? Who is going to make the food? Where will you get the food? Can participants get a quick snack and caffeinated beverages as needed? Is everyone comfortable with alcohol and if so, what do they drink?

The best estimate is based on a per diem for the host city. For example, Austin is $38 a day using a conservative corporate rate, so the total there would be about $950.

Fun

Is there anything nearby for some fun time out? It's important to let the team spend time together having fun. The book sprint is intended to be a fun event that encourages more collaborative efforts long after the intense week is done. The event can be as simple as a cookout or a hike for some sightseeing. Be sure to incorporate fun into the plan. It's important to allow time for relationship building.

Book sprint budget

For the Austin-based book sprint, the costs totaled about $,5000. We raised money by asking for donations from each of the participating organizations (RedHat, OLPC, and Sugar-Labs), and FLOSS Manuals covered the remaining cost.

The total cost is comparable to hiring a contract technical writer for a specific deliverable, yet the results from the collaborative authoring effort generated more interest in the projects, and the quality was quite high.

Why hold a face-to-face meeting?

In an article published in STC *Intercom* in January 2009 titled "Embrace the 'Un' – When the Community Runs the Event"[84], Janet Swisher relates a story about the One Laptop per Child and SugarLabs book sprint in which she was a participant. While editing, she came across a wiki page that contained an embedded note saying "confirm with Walter" – Walter Bender, founding member of the MIT Media Lab and a fellow book sprint writer who has unique first-hand knowledge and experience with the XO laptop created by OLPC. Janet walked over to Walter and asked the question that was listed on the wiki page. Without the in-person nature of the book sprint, this interaction might have taken a day or more via email.

Although extroverts prefer an in-person meeting for authoring documentation, there are also good business reasons for enhancing the collaboration and communication by holding an in-person event:

- Forging relationships that will help with ongoing writing projects and information seeking

- Keeping people motivated to continue contributing to the project

- Offering a strategy for updating and maintaining the documentation because a community has been created around both the event and the content itself

The community feeling that you build during a book sprint can generate momentum to carry the project forward, giving people the urge to work on the book in coming weeks and leaving them with the feeling they'd like to work together again in the future. It also creates evangelists for the projects and generates excitement.

Remote contributions to a book sprint

Remote participation in a book sprint can occur, but it is supplemental to the actual event. In the case of the August 2008 FLOSS Manuals book sprint, we had a daily conference call so that remote participants could report their status and any blocks to their writing. We had a constant communication channel open on IRC (Internet Relay Chat). The founder of FLOSS Manuals, Adam Hyde, was always available for technical questions in the sprint room, plus the main FLOSS Manuals developer Aleksandar Erkalovic was at the ready in Croatia to fix or explain any issues we found while sprinting.

The experience of the FLOSS Manuals sprints is that the majority of the work occurs in the "real space" of the sprint. There have been exceptions to this. The *Introduction to the Command Line* sprint held in collaboration with the Free Software Foundation is an example where there were few real-space participants and many remote participants. Whichever way the participation is split, remote contributors are very important, and it's absolutely necessary to devise a plan to coordinate the real-space and remote contributors.

There are excellent tools for assisting with communication for remote contributors. We have used phone conferences (either Voice Over Internet Protocol (VOIP) or "real" phones), but in general these are not as effective for ongoing collaboration as text-based communication. A persistently open text-based chat is good for "ambient" remote communication and better suited for collaboration with remote participants over a day, a week, or longer. Voice chat seems to put too much focus on immediate activities, and there are often dreary moments when participants are trying to think of something to say.

However, quick conference calls help the real-space book sprint team remember that they are not working in isolation and that active, real people somewhere else on the planet care about what they are doing and want to help. They can give a motivating boost but are not absolutely necessary to coordinate work.

FLOSS Manuals provides the following tools for online communication among contributors: a chat widget online, IRC, email notifications of changes to a topic, and a comparison page showing the changes from one contribution to the next.

Online chat widget

To chat with others while contributing to a book, we have provided an inline chat window. You can see it on the right-hand side of all pages in the WRITE section of FLOSS Manuals website and also on the editing pages. It is a simple mechanism that works in your browser and requires no extra plug-ins.

The chat works by sending messages to the FLOSS Manuals IRC (Internet Relay Chat) channel. The text format is short and there is no history or ability to scroll back. The FLOSS Manuals chat is one chat room; there is no provision (yet) for chatting on a book-specific basis. All comments are out in the open for all to see and remain there until someone else starts a discussion and forces your comments of the page.

For more elaborate conversations that you want to record, or for private messaging, you should use an IRC client. There are many to choose from, and the best option is probably to use whichever tool the other participants and subject matter experts are using. For other projects, there may be a preferred enterprise instant messaging tool or a real-time chat or video capability that you can use. Figure 3.1 shows the FLOSS Manuals chat widget in action while someone is editing a chapter.

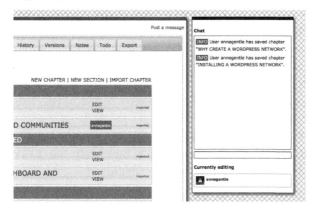

Figure 3.1. FLOSS Manuals chat widget

IRC (Internet Relay Chat)

IRC is an "old school" technology. Its "Channels" are mainly inhabited by geeks, especially free-software geeks. For chat, it's a good technology because it requires very little bandwidth: if you are on a slow connection it works quite well. IRC chat rooms or channels are contained within "IRC Networks." FLOSS Manuals uses the Freenode IRC network, one of the oldest and most popular. Anyone can create a chat room on the Freenode network, and the FLOSS Manuals chat room most frequently used is #flossmanuals. The web-based chat mentioned above connects to that main channel.

To use the Freenode IRC chat rooms, you need to have an IRC client or a browser plug-in installed. The server is at irc.freenode.net. After you connect to the server, you can join a chat room. There are other real-time chat options available such as Campfire and Etherpad. The persistent, synchronous conversations and discussions groups can have with these tools are extremely valuable when pulling off a book sprint.

Notifications

Every book in FLOSS Manuals has an email notification feature and live visual notifications of who is editing. By subscribing to email notifications, you get email every time a change is made in that book. These notifications also link to a "diff" view that shows the recent changes and highlights the differences. Both types of notifications are great during book sprints for keeping an eye on who is doing what. If someone changes something you have written, you can see the change and contact them if necessary to discuss the change.

In the editing interface (Figure 3.2) you can see who is editing any chapter. This view is useful for seeing how much activity is going on, and it also helps coordinate contributions.

TYPE ON THE GRID	edit	needs layout	TypeOnTheGrid	
COLOR THEORY	edit	needs layout	ColorTheoryAndBasicShapes	
LINE ART & FLAT GRAPHICS	edit	unpublished	LineArtAndFlatGraphics	MelChua
FIREFOX				
SEARCHING AND SAMPLING	edit	unpublished	Searching	
GIMP				
SCANNING	edit	needs layout	Scanning	
TONAL SCALE	edit	needs images	TonalScale	ChristopherBlount
LAYERING AND COLLAGE	edit	unpublished	LayeringAndCollage	DaveMandl
REPETITION AND CLONING	edit	unpublished	RepetitionAndCloning	JenniferDopazo
NON-DESTRUCTIVE EDITING	edit	unpublished	NonDestructiveEditing	
GRAPHICS FOR THE WEB	edit	unpublished	DigitalOutputForTheWeb	
SCRIBUS				
MULTIPLES: CREATING OF UNITY	edit	needs layout	MultiplePagesUnity	
MULTIPLES: CREATING TENSION	edit	unpublished	MultiplePagesChaosTensionDisarray	PatrickDavison
KOMPOZER				
HELLO WORLD	edit	to be proofed	HelloWorld	
FILES AND SERVERS	edit	to be proofed	FilesAndServers	
SEPARATING FORM AND CONTENT	edit	unpublished	FormAndContentOnTheWeb	AdamHyde
PROCESSING				

Figure 3.2. Editing notifications

Book sprints outside free software

With communication tools assisting the authoring and an authoring tool that enables quick authoring and edits, you can accomplish the goal of a book in five days. IBM has been holding similar in-person events with their RedBooks projects for years. They would give subject matter experts a half day of training in FrameMaker, have them collaboratively outline a RedBook, and then send them on their separate ways after a week of authoring and in-person discussions. Some RedBooks have moved content into a wiki for internal collaboration.[7]

Think of an Agile development team with an embedded writer, meaning that a team consists of developers, designers, testers, and writers. Accomplishing a user story about documentation may benefit from the advice given in this section about book sprints. The entire team may decide to use a wiki and collaborate on a book-like project for an iteration. You could use your planning day to outline the deliverable, determine the audience, and invite others to share in the fun of delivering a book in a short time.

---

[7] http://www.slideshare.net/almondjoy/redbooks-wiki-central-texas-dita-ug-presentation

4

Commenting and Connecting with Users

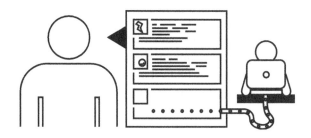

The biggest mistake is believing there is one right way to listen, to talk, to have a conversation – or a relationship.

—Deborah Tannen

It was impossible to get a conversation going, everybody was talking too much.

—Yogi Berra

New social computing tools are being invented all the time, and traditional websites are also finding ways to incorporate tagging, sharing, and other collaboration helpers in their content. Blogs have led the way. Blogs, short for weblogs, are websites with reverse chronological entries (listing them from newest to oldest). Topics range from personal to professional, offering galleries of photography or comics, and can be written, audio, or video entries.

When blogging, realize first and foremost that a conversation doesn't have to be a direct connection between writer and customer. You can blog about your area of expertise, which shows the customer your passion for your work, which translates into high quality products or services from you when representing your company on a blog. A blog entry provides an opportunity for a customer to connect to you.

Also, think about how you can help customers connect with each other in comments or in trackbacks, which notify the original blog author that you have linked to a particular entry.

However, blogging is not the only way to connect with customers or readers. You can use comment threads, online forums, Twitter, and other conversational tools. This section discusses ideas for starting conversations, building on the stages of listening, participating, and offering a platform.

Monitoring conversations

If you are a technical writer, you are probably a fast reader and collector of information, and therefore you would probably be an excellent blog monitor, able to easily handle multiple notifications from news feeds or RSS subscriptions. If you also maintain a blog and online presence, then commenting on other blogs in the field is a great way to join the conversation. Also consider joining or moderating a customer forum or board. If your product is sold on a website like Amazon, you can find conversations going on in the reviews and review comments as well as blog entries.

Reading and commenting on blogs

I believe the best starting point for blogging is to read blogs. Follow the ones that interest you personally and professionally by using an RSS feed reader such as Bloglines or Google Reader. As you begin reading and collecting subscriptions to blogs, you will notice the discussions in the comments.

Once you feel confident and interested enough in a topic or post that you can contribute a comment, write your comment, and make that connection. If beginning even an internal-audience blog is intimidating, make your first foray into blogging commenting on other's blogs. And you certainly should read blogs often before beginning to write one yourself to get a sense of style, tone, and voice.

Starting and maintaining a blog

Starting a blog may seem overwhelming at first. Not only do you have to choose a tool and platform with which to start, you must also begin the daunting task of writing and maintaining regular posts. The authors of *Groundswell*[19] say that a good measure of whether you have the stamina and energy to keep up with regular blogging is to try to write five entries. If you cannot sustain the writing effort for the time it takes to write five entries, you should reconsider starting a blog.

Choosing a blogging platform

When I started blogging, I was fortunate because the blogging tool had already been selected. All I had to do was get registered and start writing. I could also practice with the blogging interface itself, learning what I liked and disliked in a blogging engine. Because the engine choices were out of my hands in my early blog writing attempts (the internal BMC blogs were maintained on a Sun blogging platform), I did not have the added pressure of evaluating tools.

When evaluating a blogging platform, think of the authoring requirements, but also consider the reading and consuming requirements and expectations that readers bring to a blog. Here are some basic considerations for blog authoring and maintenance:

- Hosting (self-install or hosted options)
- System requirements if self-hosting
- Storage needs
- Spam controls
- Categories or tags for retrieval and subscriptions
- Access control on individual posts
- Archiving
- Trackbacks
- Post authoring and editing
- Commenting and comment notification and moderation
- Workflow for approval and timed publishing
- Multiple author features
- Content import and export
- Syndication
- Templates
- Web analytics
- Themes and styling capabilities

If the blog platform is not already selected for you, compare blog platforms at weblogmatrix.org to assist in your selection process. You can filter based on many of the considerations listed above, including system requirements, features, and support options.

An important consideration for choosing a blog platform is the ability to connect to your audience using comments and to provide link notifications called trackbacks. Trackback notifications are sent to a blogger when you link to his or her blog entry. They typically only work within the same blogging system. For example, WordPress blogs cannot notify Blogger blogs about a link to an entry.

Other important considerations include spam prevention and comment moderation. And if you need to design a theme for the blog to match your company's brand, consider how easy it is to modify the templates yourself.

The basic expectations that readers have for a blog are comments, subscriptions, and easy linking to individual entries. Without a subscription system, like an RSS or Atom feed, a website cannot be a blog.

Frequency of posts

When I first proposed starting a blog for BMC Software, my role was writing solutions documentation for combinations of products that solved specific business needs. My proposal was accepted immediately because of the type of customer we were trying to reach with our particular type of writing assignments.

Blogging about my experiences while I learned about ITIL (the IT (Information Technology) Infrastructure Library) and Business Service Management would help others learn with me. This type of "learn with me" message is a good match for the blog medium.

Before writing the proposal, I asked a fellow technical writer – who I knew was a blogger with a decent following – how many posts a week would I need to promise in order to keep a set of readers? She said two a week at a minimum with three a week being ideal.

I wrote into the proposal that I would write a minimum of two posts a week, which would take about four hours or ten percent of my time. I kept that schedule up, aiming for a Tuesday and Thursday post, and was one of the more prolific bloggers on the site.

A few months after launching talk.bmc.com, they instituted a policy that you had to have at least one blog update every two months or your blog would be taken off the site. This type of requirement is a good idea for corporate sites that want committed bloggers and dynamic content.

My initial work in blogging was internal to BMC Software. This limited my potential readers, but it helped me focus my audience. I could practice my voice and write limited topics while finding what I liked about blogging and where the rewards might be so that I could justify the time and effort I was investing. I learned about post and commenting frequency, where comments came from, how to respond, and how to grow readership.

After getting comfortable with and learning from my internal blog experience, I next began blogging externally on the talk.bmc.com site, using the Plone engine already set up for us. Because comments were moderated by other BMC staffers, I did not have to worry about deleting spam or moderating comments; I could focus on researching and writing entries.

If your company offers any blogging platform, I highly recommend that you use it as an experimental path for learning about blogging and determining if there is a return on investment (or "Reach and Influence") equation that will work for your blogging efforts.

Continuity of posts

If you want to take a break from blogging but still want to have fresh blog posts on your site, you could write posts ahead of time and set them for a publish date in the future. If you can't take the time to get that far ahead in publishing you can also ask for guest posts. This technique works well for planned extended leaves such as maternity leave or for avoiding burnout.

In my case, when I was out raising babies, I wanted the voice to be genuine and current, so I chose not to pre-write posts. Instead, in planning for my leave, I wrote to about twenty of my colleagues and asked each of them to write a blog entry about a specific topic, selecting topics that I knew were interesting to them or that they had written email messages about previously.

I managed to get about ten posts this way. I introduced each post with a short introduction to the guest blogger and a note to let the reader know that I was on an extended leave. The supportive talk.bmc.com team published them for me on a weekly schedule. This technique proved to be an effective bridge, and I did not lose readership while I was away.

Blog examples

Atlassian Confluence: A technical writer's blog

Customers seeking technical support for a Confluence product have commented on Sarah Maddox's blog shown in Figure 4.1. They know that she is a technical writer at Confluence, and they have come to expect a quicker response from her personal/professional blog than from other traditional channels. And sure enough, she delivers on their expectations.

Figure 4.1. http://ffeathers.wordpress.com

I'd Rather Be Writing

Tom Johnson has been blogging tirelessly for years now. While he doesn't speak on behalf of his employer, he often draws from work experiences to engage his readers.

Figure 4.2. http://idratherbewriting.com

Customer blog infrastructure

In many organizations, the technical publication department will not take the lead in designing or providing infrastructure for customer blogs. This type of service is often better performed by the sales department, which can track generated leads more easily, the customer support department, which has a business goal of customers helping other customers, or by departments like website management or marketing. Sometimes a valuable internal communications device is started independently by someone using a server under a desk. For inspiration, read *Groundswell*[19] for case studies such as the Best Buy internal blog platform for employees.

Integrating user content into user assistance

"Mashup" typically refers to websites or applications that combine data from different sources in new ways. Integrating user contributions into your user assistance could be considered a mashup. Another example of a mashup that could be used in user assistance is a "wikislice," which repurposes part of a wiki as an ebook or online help system. A wikislice is a cross-section of a wiki created for a particular audience.

The sample wikislices on Wikipedia[5] contain content related to mathematics, chemistry, and physics, to name a few subjects. A wikislice can be downloaded for use offline, so an Internet connection is not necessary once the content is local. And because the wikislice can be styled differently from the wiki as a whole, it can be given a format more appropriate to its particular content.

A project called InfoSlicer[6] (Figure 4.3) is housed on the SugarLabs website, host site for the developers of the Sugar learning platform. A team created InfoSlicer to help teachers create curriculum by downloading and re-mixing Wikipedia articles. For example, if a teacher wants to teach students about the African continent, he or she can download articles about animals, clothing, civil structures, and traditions that are native to Africa, then combine them into one set of articles. InfoSlicer uses the Darwin Information Typing Architecture (DITA) map structure to combine the data.

---

[5] http://en.wikipedia.org/wiki/Wikipedia:WikiProject_Wikislice

[6] http://activities.sugarlabs.org/en-US/sugar/addon/4042

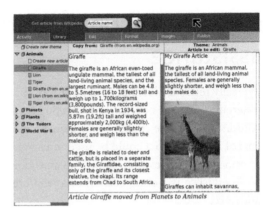

Figure 4.3. InfoSlicer for creating wikislices

To use some slice of a wiki as part of your online help system, you could set up a plug-in that queries the wiki's database or source content for specific articles at specific publishing times. You need to ensure that the community would welcome such remixes of their contributed content and that the wiki content licensing allows you to republish the content as part of your static user assistance system.

If you are serious about producing context-sensitive help using a wiki, investigate Sarah Maddox's book, *Confluence, Tech Comm, Chocolate: A wiki as platform extraordinaire for technical communication*[20]. She has a case study about web-based, context-sensitive online help that describes in detail how to create the hooks for context-sensitive help calls from within a software application.

Following are tools and techniques that can enable a mashup of wiki and user assistance:

- MWDumper, available at http://download.wikimedia.org/tools, is a tool that extracts sets of pages from a MediaWiki dump file. You write filters to select the pages you want. Your information architecture should expect to export the latest version of a page and you can also limit the export to certain namespaces. This filtering enables you to pre-select an entire namespace (removing the discussion or "talk" pages) to be filtered and exported from MediaWiki. Prerequisites are the Java SDK and SQL access to the database running the MediaWiki installation.

- You can use a Special:Export box for MediaWiki pages that enables you to export a set of pages based on a category name. On any MediaWiki installation, you can enter the root domain URL plus "/Special:Export" to get an XML export of a page. For example, go to http://www.wikihow.com/Special:Export and enter `Work-World` as the category. You'll get a list of the subcategories. Select the options and save the file. You can then download the resulting XML file.

- If the structure of the content is easily determined (such as a heading followed by only paragraph text), you might consider screen-scraping, which can be done with scripting programs. For more complex topics, such as tasks with embedded graphics, this approach may not work well because it is difficult to capture multiple embedded files with a screen scraping process.

- You can also directly import into the wiki "source" layer using a tool like DITA2wiki,[7] which is available on SourceForge. DITA2wiki is a toolkit that enables you to publish DITA content (maps and topics) to a wiki, the Confluence wiki in particular, but it is an extensible framework.

If you're thinking about mashing up your wiki content with your user assistance content, consider the following guidelines:

- Ensure that the content is licensed for use in a new deliverable. Creative Commons licenses are commonly used for works that are meant to be shared. There are several varieties of Creative Commons license, some of which allow you to create derivative works like mashups.

---

[7] http://sourceforge.net/projects/dita2wiki/

Some varieties of the Creative Commons license, as well as the GNU Public License, permit modification, but require you to license any derivative content under the same conditions as the content you started with. These licenses typically also require you to attribute the original source of the material.

 It is a good idea to get legal advice to help you determine the best license for your material and to make sure you understand the license on any material you want to reuse.

- Design the wiki with specific tags in mind for collection. Let wiki contributors know that articles tagged in a certain way will be incorporated into the online help.

- Query the database for categories, specific tags, or templates, and create a DITA map or similar structure for reassembling the content.

- Using a DITA map or similar structure as a query, get the content out of the wiki database, and then turn it into HTML files (or whatever format is needed by the user assistance/ help authoring tool). Import files into the help authoring tool and recompile the help.

Other ways to integrate user content into your user assistance include using screencasts created by users, searching Google for the most helpful troubleshooting articles related to your product and asking permission to re-post those, and incorporating wiki articles with the author's permission and proper licensing agreements in place. These suggestions are just a few techniques that might work for your content.

Reaching out to bloggers

One technique I have employed in my work with OpenStack is to find ways to highlight bloggers in the community and build relationships with them, inviting them to contribute their blog entries to the documentation. This approach works particularly well when the bloggers purposely license their blog entries and images with Creative Commons licensing. Maintain Google Alerts for blog entries that contain particular keywords related to your documentation. Then introduce yourself and reach out when a blog entry looks relevant to your audiences and could be integrated into your documentation. Also, to borrow a page from Sarah Maddox's excellent play book, you can maintain a curated page that points to helpful blog entries to compensate bloggers for their efforts.

Introducing comment and feedback systems in user assistance

Help authoring tools can provide user feedback mechanisms inside your topics. Even simply providing an email address link on each help page will give your readers a way to give feedback. If you want to structure the feedback, add a web form with specific fields to be filled out before a reader can send it. The feedback mechanism can send the form via email, which is a simple starting point. If you like to do regular surveys of your audience, Google Forms can populate a Google Spreadsheet with results even as the responses are still coming in.

Third-party comment and feedback systems can be incorporated into many online help systems. Jive Software, Disqus, and ECHO are examples of such offerings. See the section titled "Selecting and implementing a commenting system" (p. 80) for more information on selecting a commenting system.

Regardless of how you collect comments, make sure you track which page each comment comes from. Textual analysis and pattern detection can also be used to help you shape your documentation. And you may find ways to discover what your readers are doing without requiring their participation to do so. One example of this can be found in Will Hills "read-wear" phrase and concept discussed in Chapter 5 (p. 91).

Selecting and implementing a commenting system

When selecting any type of documentation system, you should consider who will use the system. I use personas as part of this exercise. As you gather requirements for your commenting system, you need to balance the needs of the different personas, recognizing that some personas are going to be better served than others, and that you will need to make decisions based on your assessment of what each persona needs. Here are the four personas I would consider when selecting a separate comment system to integrate into your documentation system:

- **Commenters:** People who expose their identity to the system and communicate in reply-based discussions. They may enter a lot of text to get their point across. They probably want to be notified when responses appear and to be able to choose whether to subscribe to ongoing comments. They may need to comment from a mobile device. They may need to paste in snippets of XML or HTML code.

- **Readers:** People who read the documentation and comments to gain a deeper understanding or to troubleshoot a similar problem. Readers may also subscribe to comment threads and search past comments. They may need to read comments on a mobile device.

- **Moderators:** People who read all the comments, respond to comments, get notifications on new comments and replies, and look for patterns in the comments. Their role is to be responsive to comments, building a sense of trust that someone is listening. This person also looks out for spam, fake identities, and useless or harmful comments. This person configures the system for comment moderation and decides whether to hold or publish comments.

- **Administrators:** People who may be managing comment threads on several sites, who need to do bulk actions on comments or pages containing comments, and who are looking for patterns in comment threads and analytics on conversations and connections.

I have purposely left out the "author" role as a persona because I believe authors should step into all these roles. As the original page author, you may need to fulfill each role in turn. While considering these personas' needs, also ensure that you gather requirements from teammates and coworkers and community members who will fill these roles for your commenting system.

Here are some basic system requirements for a commenting system:

- **Identity:** Consider integrating with your company's identity system or with social sites that provide an identity, for example, Facebook, LinkedIn, or Github. Think about the type of community you want to build and advocate for. Are they like the "social coders" on Github, or are they consultants looking for business on LinkedIn? The community you want to build may already have an identity system that they trust in a separate social networking site. In these cases, you want to build identity management through that system.

- **Connection to networking and sharing sites.** In addition to commenting, people may want to share their comments in hopes of helping others. Plus they may want to share the page itself. Look for commenting systems that enable content sharing through the sites you expect your doc community members to hang out on. However, beware of this feature when trying to maintain a "beta" or partially private site where you only want members to see other member's comments.

- **Reply/Authoring experience:** Be aware of restrictions while entering and editing comments. For example, some commenting systems have a strict rule that code snippets are not allowed in a comment for security reasons. When the audience you are trying to help with your documentation needs to input code snippets so you can troubleshoot, this limitation can cause problems. Other systems have difficulties when you input specific capitalization, such as CamelCase entries, and especially when supporting code or Linux systems, correct capitalization is crucial. But these requirements depend on the types of comments you expect. For the most part, you want a commenting system that doesn't "get in the way" and that is easy to use.

- **Spam protection:** Spam filtering is crucial to building a community that trusts that your comments come from people who are asking and answering questions, rather than selling a product.

- **Expiration of comments:** You may find that your community needs a "request for input" time period on a document, after which comments are closed and threads marked as expired. When you "close" comments, the comments remain, but no new ones can be added. We have seen this need in specification documents, where the author puts out a call for comments during a specific time period, then implements the specification based on the input. Closing or expiring comments is different from deleting comments, so be sure you adopt a system that understands this nuance.

- **Bulk operations:** With an API in place for the commenting system, it should be straightforward to delete or close an entire set of comments, for example, when a new release of software comes out and the comments on the old release site should be deleted or closed. My instinct says it's best not to delete comments ever. However, closing the threads in bulk stops the discussions.

- **Notification:**   Notification options help people control the number and type of notifications they receive. Email and RSS feeds are common, with text messages (SMS) being less common. Disqus and other systems can be configured to post comments on social networks like Twitter and Facebook. This notification may be especially difficult to manage in a "closed beta" period when you do not want customers to tweet their comments and link to a private URL.

- **Analytics:** Paying attention to pages that cause a lot of discussion or questions can tell you when a particular topic causes confusion or has other problems. Analyzing comment patterns should help you grow your community and discover patterns of interest or questions. For example, what time of day are comments most likely to come in? How long does it take someone to respond to a comment? How many comments are on a particular page (ask, is it hotly-debated content or simply wrong?)

- **Moderation:** Comments can be allowed to go straight through with or without a login or you can choose to hold them in a moderation queue. Your considerations here are related to the business goals for community building activities. You may want to respond privately in some cases, although in nearly all commenting systems I've used, manual moderation has not been necessary. I have seen the need for community members to "flag" a comment as inappropriate, another nice feature to have.

- **Comment placement:** Consider where comments should be displayed in relationship to your content. Some systems only enable comments at the bottom of the page, others have annotations or comments in a side column. Also think about how your content is "chunked" – is it topic-oriented, time-sensitive, or is it meaningful to allow comments on each paragraph? Most systems have one comment thread per page, but you can chunk more finely. For an example of per-paragraph commenting, see *The Django Book*[13].

Processing comments

Here are a few things to consider when you are processing and responding to comments. You probably want to receive notifications via email, so you can see comments from overnight when you start your day. You can build some efficiencies into your notifications by setting up email rules and filters. I have found that the busiest time for comments is near a release, especially just after a release. When a document is new, but the product is not newly released, you can expect a spike while readers try out the new procedures.

Time your responses so that others have a chance to answer also. I try to wait for a short time to give another community member a chance to answer. In other situations, you may want to require a response within some time period (for example, 24 hours), even if the response is incomplete. This lets the commenter know you have read the comment and are trying to find an answer.

In a year of monitoring technical documentation comments, I have noticed several categories of comments consistently recur. These are the types of comments that we see over and over again. I suggest you have some basic procedures for responding to the following categories of comments:

- **Typo or minor edit suggestions:** Log a documentation bug report and tell the commenter about the report.

- **Conceptual questions:** These are questions like, "Why does it work this way?" Give the commenter the additional information he or she needs. Then, if you think the question is of general interest, log a documentation bug report suggesting a revision.

- **Troubleshooting or help requests:** If you can, provide help directly in the comments. If not, direct the commenter to another place to receive help, perhaps support. Take ownership and follow through to make sure the problem is resolved.

- **Feature requests:** Let the user know the current status if this is a feature that is in the works or has been rejected. If the feature doesn't yet exist, let the user know that. If there is another system where the user can request the feature, either redirect the comment there or let the user know about that system.

Avoid using obvious templates in your responses. You want to give a personal feeling to your message and identify who you are and what your role is. Your online identity may do this for you, but it is best to be very clear about your role as a comment responder.

What about "emergency" comments that are desperate pleas for help? Sometimes redirection to a second support site will just frustrate the commenter, so try to show empathy for their situation in your response and carefully word it so you don't add fuel to the fire or pay negativity and overreaction forward to a support team. Be calm, direct, and helpful.

What about extreme negativity in responses? You have to respond to negativity immediately so that the person, and anyone else reading the comment, knows you are listening. Often negative commenters just want to be heard. But perhaps there is truth to the claim that "your docs suck" – if that is the case, be sure you have an answer about why the docs suck and what you're doing to improve the situation.

You may notice I didn't even put "spam" as a category. Spam comments are typically easy to detect, and most systems give you the ability to remove those quickly. With systems that require logins, those comments are so few and far between that spam moderation is rarely needed, and it's usually just a single click to mark a comment as spam. Enabling other readers to mark a comment as spam or inappropriate also helps in this area.

Moderating or participating in online forums

Professional writers who have mastered the product they document are natural respondents to online forum questions. The best approach is to read the forum regularly for a while to get a sense of the community that exists, how often people post, the amount of time it typically takes for questions to be answered, and the key contributors to the site. After you gain a feel for the appropriate approach, style, and tone for your messages, start by answering questions, being sure to disclose your employer.

Consider becoming a moderator only after you have been a participant for a while. Most online communities do not recognize any sense of entitlement that you may have because of your employer. Instead, you must earn the community's trust and offer real help, even if you only provide links to your online help. Teaching the community to fish (for information) feeds them longer than just answering questions without citing how you learned the information yourself.

Instant messaging and responses

You have probably noticed that many websites now contain "Live Help" links. If your corporate site includes such links, you might find out how the live help is staffed and volunteer for an hour, or longer, exchanging instant messages with real customers.

Interesting examples exist of scripted response programs written for frequently asked questions, using artificial intelligence stored in knowledge reserves, dictionaries, or other reference materials such as concordances to guess at a needed response. For example, IRC has a helper robot called IRCza that is a natural language artificial intelligence chat robot for IRC.[8] AOL Instant Messenger offers AIMbots, which can tell you when movies are going to show.

Artificial Intelligence Markup Language (AIML)[9] is an XML-based language that facilitates the creation of "chatbots" with various personalities and kinds of knowledge. By connecting the chatbot to a support knowledge base of frequent responses, you might be able to offer a virtual conversation with responses to a customer.

Writing reviews

Social shopping involves researching reviews and community member recommendations before or while making a purchase. For example, you can search on Amazon.com using keywords that your users might use to get a sense of what they might be reading, and who else is reading and reviewing those books. This can give you an idea of what problems they're trying to solve. To encourage and participate in social shopping, you could can also read and review books that are related to the types of problems your customers solve with your product.

Integrating social tagging

You may want to be a tag maintainer for your product, offering tag sets related to your product or service on social bookmarking and tagging sites such as digg.com and delicious.com. Keep an eye on community members who use your products and observe what tags they use as well.

---

[8] http://timberfrog.com/icqza/irczafaq.html

[9] "Using Instant Messaging to Provide an Intelligent Learning Environment," http://-www.alicebot.org/TR/2001/WD-aiml/

Tagging involves a skill that many technical writers have – determining which keywords best describe a link or image for retrieval later. Social bookmarking sites use tagging to help users group related links for later use. The social aspect of these sites is that they make tags and the content they point to visible to others.

There are literally hundreds of social bookmarking or tagging sites, but writerriver.com, digg.com, and delicious.com are particularly useful to technical communicators.

Other potentially useful social bookmarking sites include reddit.com and stumbleupon.com.

Sharing photos and videos for explanation or assistance

At sites such as Screencast.com and IgniteCAST, users can use a desktop tool to capture screenshots or screencasts.[10] The website hosts the image or video files for playback. Photos are often the best solution for showing sequential instructions rather than writing them out or using complicated 3D modeling software.

For example, a Lego fan posted instructions for building a refrigerator (see Figure 4.4) with Legos by laying out the Lego pieces, taking photos, and then adding some text for instructions (see Figure 4.5 for a photograph of the completed refrigerator).

---

[10] Screencasts are videos that show a computer screen being used to perform some task, along with narration.

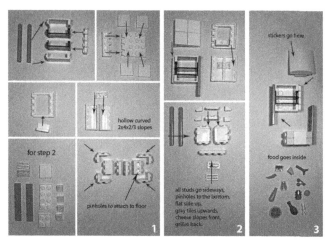

Figure 4.4. Instructions for building a Lego refrigerator[11]

Figure 4.5. Photo of Lego refrigerator[12]

---

[11] http://www.flickr.com/photos/nolnet/3184800280/in/photostream/

[12] http://www.flickr.com/photos/28119014@N06/3185664328

5

Wikis as Documentation Systems

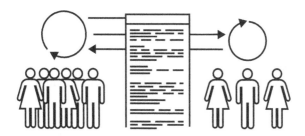

Your wiki is not Wikipedia.

—Stewart Mader

The recent explosion of wikis hosted on the Internet and the success of Wikipedia, the largest online reference encyclopedia, has led to discussions about where a wiki might fit into technical documentation. Originally created in 1995 for making quick web pages, wiki means "quick" in Hawaiian and is now synonymous with collaborative authoring. Over 130 wiki applications are available, and most offer cross-platform, cross-browser pages. Wikis encourage collaboration because they allow anyone with the right permissions to edit any web page, and many wikis can even reconcile differences when two authors edit the same article at the same time.

Wikis represent a shift towards living, breathing, changing documentation. While not all writers are certain of their value or place in technical deliverables, I believe wikis encourage crowd-sourcing – delegating work to a large group of semi-organized individuals (a crowd) – and help writers see customers' view points.

Customers may prefer a wiki when a customer forum does not provide enough information to answer their questions or when scrolling through multiple answers on a forum becomes tedious. Many of today's wikis resemble web-based content management systems, and the editors are more complex than the simple wikitext-style editing seen in the first wikis.

Wikis for projects

Wikis are good for project management and communication, for information gathering and collection of source material, and for content delivery. Wikis can be either the source of your content or the published deliverable from another source. This chapter talks about both sourcing models and also discusses round tripping, where content goes from a source file to a wiki and back in a cycle that can be repeated again and again.

Using wikis for project collaboration is an important model and one that is prevalent in today's workplace. Scott Abel, creator of The Content Wrangler, and Stewart Mader, author of *Wikipatterns*[21] and wiki adoption consultant, collaborated on research into wiki use in the enterprise, producing a report titled "Why Businesses Don't Collaborate"[39]. Their report discusses enterprise wikis, which are wikis behind a corporate firewall. The security layer for this type of wiki is based on the credentials already established for company email and the company network, making them secure and easy to access for employees. By using the enterprise wiki, rather than email, for tasks like distributing meeting agendas and posting meeting minutes, people save time, work together better, and distribute updates and information more easily. One responder to the survey noted, "I've found that people are much more likely to participate when they can just go update wiki pages vs. reviewing formal/traditional documents."

While powerful, wikis are not the best fit for every application. Some disadvantages for wikis include:

- **Portability:** Wikis are often run from a server, and even though web servers are becoming smaller and smaller, gathering up all the pieces to create a portable wiki may not be practical.

- **Connectivity:** Wikis are often only accessible when you're online, so people cannot browse a wiki on an airplane, for example.

- **Access control:** People often expect any wiki to be open for editing. If you want to use a wiki just to enable comments, you might not meet some users' expectations.

- **Output deliverables:** Many wikis offer output formats ranging from PDF to content for a mobile smartphone, but those may not meet your users' standards for output quality.

Why wiki?

When looking at wikis, make sure you understand your objectives. It may be obvious in hindsight that you need a wiki to meet your business goals, but it may be difficult to state in the first place. Ask and answer "why?" until you are about five levels down, and then it's likely you will have a solution. Here are two examples:

- Why do you need a wiki? To collaborate with customers. Why do you want to collaborate with customers? Because they have good ideas that we want to ensure get attention. Why do their ideas need attention? So that readers see those ideas and envision their own business operating in the same way. Why do you want customers to get a vision that they share with other readers? So that your product's adoption rate will increase.

■ Why do you need a wiki? To enable more writers internal to the company. Why can't the writers use the current tools? Because the current "white coat" system is too hard to learn, and it's too expensive to invest in more licenses for ten times as many writers. Why do you need more internal writers? Because a single writer cannot know enough about a complex product to write it all down. Why can't a single writer keep up with all the information? Because the pace of change is too fast. Why is the pace of change so fast? Because in a web-based environment, the speed of change is rapid.

In certain environments, a wiki may be the only way to work effectively. Your solution may or may not use a wiki, but if you're producing web content, readers have expectations about the content – for example, they probably expect it to be accurate up to the minute – that may be hard to meet with other tools.

Be aware that information stored in a wiki is easier to maintain if it's written a page or an article at a time. A page at a time is really tough for ongoing maintenance, but it's the default for most wiki engines. There are also difficulties with the rather immature technology in many wikis. Wikis were designed for simple editing and fast publishing. What if you need to change a name across 200 wiki pages? Anyone doing tech comm in a wiki knows that's a headache for many wiki systems. How about search and replace? Page at a time. What about a final spell check? Page at a time. And let's not talk about adding an entire column to a table using ASCII-based wikitext. Your wrists will revolt, I assure you.

I believe that many wikis are in the same general category as web-based "content management systems" or are moving in that direction. But there are difficulties using wikis for content management. Here are some I've heard about from fellow technical writers:

- **Access control:** Wikis that cannot control who can view or edit individual pages will not work for technical documentation. For example, without page-level access control, you can't allow editing for some articles but prevent it for others, for example, customer support articles.

- **Hierarchy:** Without at least two levels of hierarchy, technical writers will be stymied. We usually have complex documentation sets to maintain, and hierarchy and section nesting are natural ways to organize topics.

- **Version control:** Without version control, maintaining or tracking several versions of a collection of topics (or an entire namespace/space) to correlate with a software release version is frustrating, if not impossible.

- **Global search and replace plus global spell check:** Writers are used to maintaining giant, long-form documents with tools that allow them to spell check and search and replace across large amounts of content. CMSes and wikis with chunked content can make global actions more difficult to perform than tools that create long-form documents.

- **Search:** Readers and writers have become so spoiled by Google's search algorithms that most local search engines come up short.

- **Workflow:** Wikis can be weak in workflow, and even a publishing workflow as seemingly simple as "approve or reject" a particular article is beyond the capabilities of many wikis.

- **Creating or curating collections:** Wikis can be weak in capabilities that allow you to use single-source methodologies to create collections of articles based on tags, categories, labels, etc.

- **Offline access:** Many wikis only work online, but what if your users need to access the wiki offline, for example on a plane? One clever solution to this problem is to offer the entire wiki – engine and content – bundled on a USB stick: call it a wikiscicle.[1]

- **Round tripping:** Writers often talk about round tripping content – writing in XML or on the wiki and exporting/importing in either direction. I have usually dismissed this as not worth the trouble because there wouldn't be too many contributions for a team of writers to keep up with.

However, I finally heard of an example where a decent business case was made for doing this. The round trip was from structured XML (DITA) contained in a CMS to a wiki and back again. The keys to making the business case were the need for translation to 22 languages and the volume of edits and contributions.

- **One-click publishing (batch processing):** On release day, you want to set all topics to be released at once, but with many wikis, you have to go to each page one-by-one and click them over to the right state for release.

- **Wikitext:** Many wikis use *wikitext* – a simplified markup language that is an alternative to HTML – for editing. While much simpler to use than HTML, there are also some pitfalls to avoid when using wikitext instead of traditional HTML markup. For example, wikitext is not exactly the same for every engine, so learning one wikitext syntax is not transferable to another wiki.

With plugins and advanced wiki engines, these hurdles can be overcome, but knowing about these pitfalls can help you avoid them.

---

[1] http://ask.metafilter.com/71979/First-MediaWiki-now-Deki-Wiki

Comparing wikis to help systems

In March and April of 2008, 493 people responded to a survey from the Center for Information Development Management[2] that asked about publishing practices and information management techniques. More than half of the respondents said they use wikis internally, but just 9% use wikis for externally facing documents.

My interpretation is that we are getting more practice using wikis for project management and internal collaboration, but we are not yet replacing traditional help systems with wikis. However, over the last few years there has been a trend towards replacing traditional help authoring tools with wikis or with collaborative web-based content management systems. Existing, traditional tools are getting older, and the high cost-per-seat to license many of these tools has driven stretched budgets over their limits.

Wiki applications offer advantages over other HTML publication methods. Over the years watching the maturity of help deliverables, we have seen a growing demand for user assistance systems designed to be cross-platform and network friendly.

Here are some ways that wikis are different from traditional user assistance systems:

- **Editing:** The "wikitext" markup used by many wikis is meant to be simple and easy to use compared with HTML or other markup systems

- **Linking:** Many wikis allow page-name-based linking to other wiki pages, or simple HTML linking methods

- **Subscription options:** Many wikis offer RSS feeds for recent changes, and "watch" pages for edits

---

[2] http://www.cidmblog.com/?q=node/33

- **Searching:** Search engine technology often is included and indexing is done as soon as a page is added

- **Scalability:** Wikis can support many pages edited by many authors

- **Access control:** Registration and logins can control authoring and viewing privileges per page or per group of pages

Starting or reinvigorating a wiki

A wiki is not always the answer to the question being asked. Sometimes starting a wiki is just a procrastination attempt to put off the real work of collaborating and cooperating with each other. By distracting yourself with wiki application research, selection, and installation, you delay the inevitable "it's a people problem not a tool problem" realization.

The hardest part about working on a wiki is not writing the content but building the trust and relationships that will ensure the wiki continues to grow and thrive. So if you are ready to work on growing wiki use, and not just installing wiki software, here are some best practices for building your wiki.

Conduct an audience analysis

Just like with other information deliverables, knowing your audience is an important part of running a wiki. Each community shapes its wiki to match its needs. For example, WoWwiki.com is a popular wiki run by the community for the game World of Warcraft. This community wants strategy articles and reference information; users' technical or how-to questions or discussions are roundly dismissed.[4]

---

[4] Based on my notes from a 2008 SXSWi panel, "Edit Me: How Gamers are Adopting the Wiki Way" http://justwriteclick.com/2008/03/16/stories-from-sxswi-2008-edit-me-how-gamers-are-adopting-the-wiki-way/

So, before starting a wiki, offer a survey or talk to customers about their expectations.

Build a wiki style guide

You may already have a style guide for your writing deliverables. For example, you may have a list of preferred action words for tasks and steps, a description of how you capitalize headings and the titles of wiki pages, or a method for standardizing URLs. I suggest you embed your wiki style guide directly on the wiki itself as a wiki page. And of course, make sure your own contributions consistently follow your style guide.

A style guide can describe which categories are helpful in your wiki and provide some of the background behind choosing those categories. For example, the OLPC Wiki has a Style Guide page[5] that helps new contributors understand the structure and organization of the wiki. The Firefox Support Knowledge Base has a comprehensive style guide[6] that also refers contributors to Sun Microsystems' style guide, *Read Me First! A Style Guide for the Computer Industry*[33].

If contributors do not follow the style guide, even with reminders, you will need to decide whether you need to edit the text to match the style guide or if the community will accept the text without strict adherence to a style guide.

Another way of defining style is determining a "voice" for your wiki. Decide ahead of time the types of information and word choices that the wiki pages reflect. Refer to Chapter 6 (p. 147) for more suggestions and style guidance.

Offer wiki how-to guides and training

Wiki tutorials give new users a chance to learn in a safe environment, one in which they can explore and experiment and not worry about

---

[5] http://wiki.laptop.org/go/Style_guide
[6] http://support.mozilla.com/en-US/kb/Style+Guide

harming the wiki. A tutorial, or just a page with some simple how-to steps, may be sufficient. Offer a "sandbox" area or single page in your wiki where people can try out the interface and practice writing the wikitext knowing that their changes are not permanent.

Another way to help users get comfortable making edits is to purposely leave typos in certain pages, giving users a chance to correct small areas before attempting to revise an article or add a new article.

A how-to guide may be helpful for describing common tasks, such as starting an editable section or starting a new page. Even an experienced wiki user might need a short how-to guide on the basics if the wiki application is unfamiliar.

If your wiki seems to be intimidating to some, or if your customer base is not already familiar with the wikitext you are using, you may want to offer webinars or training sessions to help boost people's confidence in contributing to the wiki. If your wiki is only for an internal audience, you could offer informal brown bag lunches to demonstrate how to edit the wiki or add pages.

At Quadralay, the company that develops WebWorks, they held a brown bag lunch for the entire company before rolling out their wiki, and all but one of their employees have made edits on the wiki.

Set rules for arbitration

Depending on the subject matter of your wiki, you may have controversial articles that get a lot of discussion and comments, or rounds of aggressive edits in which one contributor is pitted against another for the "truth" or being "right." When this type of standoff happens, it is best to have rules for arbitration to determine which edit is the edit that remains. Sometimes a community leader will step into this role, but sometimes your company will have to take a stand. Either way, be sure to have a written policy available on the wiki to set expectations. Experienced contributors want to know the rules of engagement and the expectations of the community.

Another approach is to allow multiple wiki articles for each topic. For example, collaborations for online textbooks do not use the one-article-per-topic model for their collaborative authoring environments. Instead, they allow multiple authors to write more than one conclusive article, such as describing how the American Civil War started. When the textbook is compiled for printing or PDF creation, a "lens," or specialized content filter, determines which article is included in the textbook. Some wiki applications will enable this filtering capability so that readers can devise their own lenses for the content, but more often, only one wiki page "wins" the battle for being the true article on the wiki.

Differentiate "warranted" content

Many wiki users are following a publishing model that clearly designates an area of the wiki that is tested, vetted, and company-backed. A common model for many groups is to have an internal wiki where drafts are published and discussion occurs, and an external wiki that houses the content that is considered "warranted."

Warranted implies that you can log a support ticket against the content or the procedure documented and expect an action. With warranted content, you set the expectations that people bring to the content.

For example, FLOSS Manuals has a "Write" area where all content remains while it is still in draft stage and a "Read" area to which wiki pages are published when the content is ready. A small group of writers with "Maintainer" privileges has the capability to move content from the "Write" area to the "Read" area.

Lisa Dyer, a development manager at Lombardi Software prior to its acquisition by IBM, uses a similar model to maintain three wikis:

- An internal-only wiki space is used by developers, product line managers, quality testers, and information developers to write, display, and discuss use cases, test plans, and other software documentation. All employees can view this content.

- Guaranteed, tested content is published to an external-facing wiki space, and only information developers have full control over editing the wiki pages. Anyone can log on to the wiki using their registered ID and leave comments on the pages, but they cannot directly edit the content.

- In a third wiki space, Lombardi's technical services employees can edit articles, and customers can view the content. However, a disclaimer on the wiki states that the content is not tested and following the instructions may not have expected results.

You can read more about their wiki model in "Using Structured Wikis in Software Engineering"[40] from WikiSym 2008 or this slide presentation, Lombardi Wikis: A model for collaborative information development.[7]

Integrate the wiki with other content

Although some people find that a wiki is sufficient for delivering the information a customer needs, a wiki alone is not sufficient for many audiences and products. If you produce a wiki in addition to other deliverables for a product, you need to integrate the wiki with the documentation set so that it looks like the rest of your documentation, follows your style requirements, and is included in search and indexing tools. If you need to set the wiki apart because it contains unvetted content, or because it contains only tutorials or specialized content, it still should follow corporate guidelines for visual style.

If a wiki's look and feel is widely different from other deliverables, the wiki may appear to be an afterthought or unneeded part of the documentation set. Tom Johnson, on idratherbewriting.com, has found that the scope of your search facility and how well it integrates

---

[7] http://www.slideshare.net/lisa.dyer/lombardi-wikis-model

with other content in your documentation set has an impact on usability. In a blog entry, "A Web 2.0 Documentation Idea Gone Wrong"[64], Tom voices his concerns about user assistance that is fragmented and in too many locations without a unifying search or theme. When users are within a stand-alone help system, they may not know that a wiki exists. Finding ways to expand the user assistance system to include the wiki or support forums in a search would help direct readers to the conversation-based content.

You may find that you can single-source to initially build a wiki. Here are a few examples: Webworks[9] publishes to wikitext from multiple types of source files. The DITA2wiki open source project offers a framework to use DITA (Darwin Information Typing Architecture) source files to populate a wiki. The Mylyn MediaWiki to DITA project,[10] transforms wikitext into DITA XML. K15t Software[11] has software to export DocBook from Confluence wiki.

Another workflow is to author in a wiki and output PDF files as books from the wiki. Wikislices are one method for extracting a book from a specific set of wiki pages. But wiki integration could also mean supplying a wiki with updated articles and then using wikislicing to pull those articles back into user assistance that is delivered in another way.

Another aspect of integrating a wiki with other content is ensuring that an online identity is associated with the wiki. For example, if a login ID is already required for access to your customer support website, perhaps your wiki could use the same login credentials, lending credibility to the wiki because only supported customers could edit the content.

---

[9] http://Webworks.com
[10] http://greensopinion.blogspot.com/2008/11/mylyn-wikitext-targets-oasis-dita.html
[11] http://www.k15t.com/display/en/Home

Ensure that your wiki has original content

Let your wiki community know that you are fully committed to the wiki by offering original articles found nowhere else. Knowing that the wiki is dynamic and original will entice readers to continue to visit it. Even more enticing would be to feature a community member's contribution on a regular basis. This best practice works well when the wiki is just a part of your overall documentation set. If your wiki is the sole source of end-user documentation, then the wiki itself is all "original" content, meaning, its content cannot be found in other publications.

Beware of spammers, and back up often

You want to constantly check the wiki for edits or comments that do not look genuine or well-meaning. You can reverse those changes if they happen. Spammers will not target your wiki for long once they understand that their efforts are futile.

You can protect your wiki from spam-based changes by requiring users who make edits or comments to solve a math sum in a web form or enter a text string from a graphic before submitting a comment. Tools such as Captcha[12] are widely used and effective. Some wikis allow you to block, redirect, or ban users or IP addresses.

Wiki vandalism is different from wiki spam. Typically a vandal's intent is to disrupt the community or annoy or distract a particular community member. Again, immediate reversal of the vandalism and backup of your wiki can often stop vandals. Vandalism is rare and is not typical for most wiki communities. An active community can also help you keep unwanted content and spam out of your wiki. Rather than expending energy preventing vandalism, work harder on building your community and encouraging contributors.

---

[12] http://www.captcha.net/

Keep content up to date

The point of having a wiki is to enable constant editing and adding of new content. Users expect wikis, more than most forms of documentation, to have timely and frequent updates, and most wiki tools offer a timeline of changes so users can easily view the latest edits and additions.

If your product's release schedule does not typically have regular updates, you may find that a wiki is not a good match for your product documentation, and you might look elsewhere for customer conversations.

In a case study[13] published on the Atlassian website, GigaSpaces technical writer Gilad David Maayan says, "we weren't updating our help pages as often as we would have liked." He moved their online help content to a wiki to gain more frequent updates, more contributors, improved searching, finding, and a better overall user experience.

Become a member of the community

If you are willing to share your professional identity with customers, you lend credibility to your intention to become part of their community. In some cases, depending on the wiki and the corporate culture, you might not want to insert yourself too much into the community that is building up.

For example, a developer of a game might be attacked if he or she tries to defend design decisions made in the game. Plus, if writing strategy articles is not part of a developer's job description, and the wiki community members prefer that the developer continue to develop features, article writing might be perceived as a waste of resources. If instead the developer makes a useful comment on a

---

[13] http://atlassian.com/software/confluence/casestudies/gigaspaces.jsp

wiki page, he or she is still contributing to the community but in a way that the community would embrace instead of attack.

Maintain categories

Many technical writers excel in creating indexes and keywords, including inferring other terms that people might want to use. With categories on a wiki, you can organize in many ways, beyond keywords. Although you do not want to over-organize a wiki and risk discouraging community contributions, you can suggest categories, tags, and keywords to a contributor just as you would to a co-worker. However be prepared to accept their preference if they don't want to use your suggestions. Ease of use and low barriers to entry may be more valuable than strict rules about consistency.

Watch recent changes

If you want to become an active contributor to a wiki, you should keep an eye on the recent changes page, either by RSS or email notification. This practice gives you an idea of what areas are being heavily revised and also what areas might be under contentious debate.

You also want to watch certain pages, such as pages that you start or edit, to see how your changes are interpreted and whether activity happens on your page. In the case of the One Laptop per Child simple user guide, I watched the wiki page that contained the user guide content and was pleasantly surprised when Walter Bender, the president of development for the project at that time, contributed edits from his wealth of experience in deploying the laptops.

Learning from Atlassian: Interview

Sarah Maddox is technical writer at Atlassian and author of *Confluence, Tech Comm, Chocolate: A wiki as platform extraordinaire for technical communication*[20].

1. **How are you aligned in the company? Support, marketing, training? What is your job title?**

   I am one of the technical writers at Atlassian. We are part of the engineering department and work closely with product management, support, and marketing, as well as the development teams. Our two CEOs have a keen interest in the documentation, too. Many of the pages flaunt one or the other of their names as original author.

2. **What types of content do you deliver?**

   Atlassian develops and sells a number of products, mostly web applications. The documentation covers them all. For each product, we supply user-focused manuals (user's guide, administrator's guide, installation guide) and developer documentation (API references, toolkits and tutorials). We also write the release notes for every product.

   Release notes are a curious beastie. They need to be technically correct and informative, appealing to look at, and just a bit flashy. We follow an iterative and intensely collaborative process to achieve such hybrid perfection. The technical writers create the framework and the first draft of the content. Product managers tweak the words, rearrange the items to reflect the key features in the release, and add screenshots with rounded corners. The marketing specialists add a video and tweak the words again. Technical writers fix the typos and ensure all is present and correct. We iterate until a minute before the publication deadline, then push the button.

   All the documentation is developed and hosted on a Confluence wiki, at http://confluence.atlassian.com. People surf in from Google and other search engines. People drop in via links from blogs and websites. People click context-sensitive help links on the application screens and arrive at the appropriate page on the wiki. Some customers cannot come to the wiki, because they are stuck behind a corporate firewall or are offline much of the time. For them we export the guides to PDF, HTML and XML, and provide the files for downloading from the wiki.

As well as the documentation itself, the technical writers review the wording of the user interfaces and error messages. That is, after all, the first help text that people see. Every now and then we write posts on the corporate blog and contribute material for other publications.

3.   **How is your content licensed and how was the license selected?**

We provide the documentation under a Creative Commons Attribution (CC BY) license. This means that people can use our content for any purpose, provided they acknowledge the source by referring to the documentation wiki. People can even copy parts of the content, adapt it, and incorporate it into their own manuals.

Community authors and developers can log in to the wiki and add pages or update existing pages in the documentation. We ask them to sign a contributor license agreement which protects the rights of the contributors, the company, and any third party whose content may be involved. Contributors also understand that their content is licensed under the CC BY license, since it applies to all the documentation on the wiki. Compiling the license agreement was an interesting exercise. We based it on the Apache Software Foundation's Contributor License Agreement (http://www.apache.org/licenses/icla.txt), which is designed for software rather than documentation. I made a draft of what I thought a documentation agreement should look like, then we engaged intellectual property lawyers to knock it into shape. It was very satisfying to help devise a solution for collaborative documentation.

4.   **How many collaborators do you work with regularly?**

Everyone in the company, nearing 400 people, has update access to the documentation. In practice, probably ten people will update the documents for a single product on a regular basis. In addition to the internal authors, we also have 23 community authors who have signed the contributor license agreement. Three or four of those update the documentation regularly.

Readers can also add comments to the pages. They do not need to sign the contributor license agreement or even log in to the wiki. Comments are a hive of activity. Sometimes they are a lot of fun. Sometimes they contain spam, and those we delete immediately. Mostly, people ask and answer questions, point out things that need fixing in the documentation, and offer each other hints about using and extending the functionality of the software.

The technical writers manage all this by setting the wiki permissions to suit each type of content and by monitoring activity via RSS feeds or wiki watches. To avoid being swamped by wiki notifications, we allow update access only on the latest versions of the documentation. Even so, first thing in the morning is a busy time as we hook ourselves up to our feeds and see what people have been doing while we were asleep.

5.   **Do you hold sprints, and if so, how many people are active in sprints?**

The engineering teams practice various forms of agile methodology, so our day-to-day documentation tasks are closely aligned to the development sprints. Every now and then, we run a special sprint to get something specific done.

1. We have held two doc sprints, and intend to do more. The first doc sprint focused on developer documentation, specifically plugin development tutorials. The second focused on user documentation. The sprints were a great success, in terms of both the documentation produced and the satisfaction felt by the participants. There were 23 participants in the first sprint and 30 in the second. The authors were from all over the world, including Russia and Israel, working remotely as well as in our Sydney and San Francisco offices. At the end of each sprint we held a retrospective, in true agile fashion, where people discussed what went well and what could have been better. The biggest lesson learned was the value of templates, essential in giving sprinters a starting point and ensuring consistency of output.

2. We have also held a couple of documentation blitz tests. This is a very successful way of involving the development and support teams in testing the documentation just before the release date. The technical writers set up a plan, including a list of the documents to focus on and a couple of ways people can give us feedback. We usually include an IRC channel, as well as wiki pages and comments, so that the engineers can choose the way that suits them best. We allocate a time period, usually just an hour, and everyone dives into the documentation. The chat session goes wild, comments fly, and we end up with a lot of useful feedback.

6.   **How many page views do your pages get?**

For the month of 22 July to 21 August 2011, our Google Analytics report shows a total number of 712,709 visits to the documentation wiki. The number of unique visitors was 406,966 and the number of pages that they looked at was 2,186,923. The report also gives other interesting figures such as the percentage of new visits (45.78%) and the average amount of time each person spent on the site (4 minutes).

7.   **What is your tool chain and how is it created? Internally, with outsourcing, through the community?**

Our primary tool is Confluence wiki. It just so happens that Atlassian is the maker of Confluence, so writing the documentation on the wiki is a dogfooding win too. ("Dogfooding" is a horrible word that is apt to make a technical writer's hackles rise. But it has become an accepted term. It originates from the phrase "to eat your own dog food," which means to use your own product so that you can personally experience its good points and its bad points.)

We use a few plugins that extend the core wiki functionality to satisfy the requirements of technical documentation. A plugin is similar to an add-on for a web browser. It is a piece of software that you can install to add functionality to the wiki. Some plugins are built by Atlassian and others by community developers. The Copy Space plugin[14] is indispensable for a documentation wiki. We use it to create a new space for each version of the documentation.

---

[14] https://plugins.atlassian.com/plugin/details/212

For technical diagrams we use the Gliffy Plugin for Confluence.[15]

We also sometimes work with plugin developers, explaining requirements of technical documentation such as workflow and version management. This collaboration becomes a joint effort to produce a plugin that will enhance the wiki, satisfy a requirement that many customers have, and build a marketable product for the plugin developer. Interesting and rewarding work indeed.

8. **Anything else you want to share about your particular situation?**

One thing that makes us fairly unique is the number of initiatives we have put in place recently to draw people into the documentation. We encourage collaboration of different sorts, both remote and on the page.

1. At each major software release, we Tweet the highlights from our release notes. Each Tweet includes a link to the release notes in the documentation.

2. We encourage people to Tweet hints and tips about our products, and we display the resulting Twitter stream live on a wiki page. This means that people can swap tips in Twitter, and they can also see their own Tweets appearing in our official documentation. An example is the "Tips via Twitter"[16] page for Atlassian JIRA.

3. There is a page called "Tips of the Trade" in the documentation for each product. The page contains a categorized list of links to blog posts

---

[15] https://plugins.atlassian.com/plugin/details/254
[16] http://confluence.atlassian.com/display/JIRA/Tips+via+Twitter

by external authors. If someone has blogged a technical tip about the product, they can add a comment to the page and we will link to their post. This is a great way of engaging bloggers in the documentation.

It is fun and rewarding to work in an open environment where technical communicators are encouraged to speak out and try something new. For customers and community authors, the documentation becomes a hub where people can swap ideas and help each other, as well as find the answers to their questions.

It is true that collaboration and social tools are not suitable for all technical documentation. Audience, environment and other requirements are paramount, as always. I have worked in a number of organizations using the full gamut of tools. Now I am lucky to be working on a wiki, and in a forward-looking company that recognizes the value of documentation and communication. Our customers and management constantly challenge us to evaluate the quality of the documentation and of our team's contribution to the company.

Expect small percentages of contributors and value them highly

As with other types of online communities, you will not have hundreds of content contributors. Realistically, you might have a core group of about five contributors for most types of information products. Several sets of research provide insight into this pattern.

Jakob Nielsen's 2006 article, "Participation inequality: Encouraging more users to participate"[71], describes the "90-9-1 rule," which claims that participation in online communities (forums, email lists, newsgroups, etc.) breaks down into about 90% lurkers, 9% small contributors, and 1% large contributors. He studied email lists, online forums, and newsgroups, which are the natural predecessors of wikis. He says these numbers have been observed for years and that you can't really change these values by much – perhaps by only a few tenths of a percent.

His article concludes with suggestions on how to move those numbers slightly, either towards more contributors or towards more lurkers and fewer elite contributors (which may also be a desired result). Nielsen's suggestions include ideas that are easily transferred to user assistance or to wiki contributions. First, make it easy to contribute quickly, for example by offering a star rating capability instead of only a web form. The ability to rate an article is included in many online help systems, and exposing the ratings to the reader helps to determine how "well-worn" a help topic is. With forum software you can also determine the value of a community member's contributions based on reputation systems.

Another suggestion is to make participation a side effect. I've explored this idea in a blog post, "Can online help show 'read wear'?"[54], that looked for ideas on how to make a user assistance system show the fine lines of wear, like a cookbook that falls open to your favorite recipe because the binding has been flattened on your kitchen countertop. This indicator of use is called "read wear"[60]. Here are some ideas for indicating "read wear" from the blog, along with ideas from some of the people who commented on the entry:

- Model a wiki's "most active pages" feature, which shows the page with the most edits. However, the page with the most edits may be more controversial than truthful, so the most popular pages might be more useful.

- When a user searches, show the most searched-for terms. Users might connect concepts more easily when they understand what others are searching for.

- Show the most recently viewed knowledge base articles or most popular articles. I found that useful in the past when searching through BMC Software's rich knowledge base. Just as CNN and other news sites offer a list of the most emailed stories per hour, you could show the most emailed online help topics.

- Use tag clouds to display read wear, with the size of the tags indicating the number of topics users have tagged.

I found that popularity, time spent on the page, rating on a page, and number of edits on a page are the strongest to weakest indicators of read wear for an online content system.

Nielsen encourages the use of templates and draft content so users don't have to start from a blank page. I know a writer who pre-filled as many technical articles with as much information as he could before sending the drafts to his Subject Matter Expert (SME). The SME, a recognized programmer who built software for electronic keyboards, happily added content into the pre-filled articles.

Nielsen also says to reward, but not over-reward, participants, and he encourages the use of a reputation system. I especially like the idea of moderate rewards for participation so that dominant profiles do not take over the community. Balance is key, along with recognizing that you won't overcome the community makeup.

Growing your community should also contribute to the growth of participation. Andy Oram's study, "Rethinking Community Documentation"[72], of why people contribute documentation to free and open source software projects shows that building communities is the clear winner.

Applying the 90-9-1 rule makes the growth of a community concrete. For example, if you start with 3 writers out of 300 wiki participants and grow the community to 3,000 wiki participants, it's possible to have 30 contributing heavily to your wiki project if the 1% large contribution number holds throughout the community's growth.

Some wiki examples show variances in these percentages, but as described earlier, certain best practices can shift the participant percentages. Following are two examples of wikis that have had their participation percentage values studied.

Wikipedia contributors

The article "Who Writes Wikipedia"[83] by Aaron Swartz discusses the claims made by Jimbo Wales that on Wikipedia, "over 50% of all the edits are done by just .7% of the users... 524 people." Aaron Swartz describes the story he saw from his studies as follows:

> When you put it all together, the story become [sic] clear: an outsider makes one edit to add a chunk of information, then insiders make several edits tweaking and reformatting it. In addition, insiders rack up thousands of edits doing things like changing the name of a category across the entire site—the kind of thing only insiders deeply care about. As a result, insiders account for the vast majority of the edits. But it's the outsiders who provide nearly all of the content.

MSDN community content contributors

Although not precisely a wiki, the Microsoft Developer Network[17] has community content features that are wiki-like. According to the site, as of December 20, 2007, 1,866 edits out of 10,851 total edits

---

[17] http://msdn2.microsoft.com/en-us/library/default.aspx

were made by the top five contributors (three of whom are Microsoft employees). That percentage is slightly above 1% at 1.72%. Looking again on April 14, 2009, there are two Microsoft employees in the top five. There were over 6,000 contributions from the top five contributors out of 31,615 total edits, but a contribution can be a comment or a topic, so it's more difficult to measure contribution percentages.

Recruit other internal reviewers

Although most online communities strictly adhere to the 90-9-1 rule (where only 1 percent of the community contributes content), you can shift those numbers slightly by recruiting others in your company to offer their perspectives to the wiki and do minor edits and upkeep when needed. Have someone from customer support keep an eye on the recent changes or comments RSS feed and also ask a marketing or sales representative to do the same.

Do not expect a wiki to fix other problems

A wiki is an open documentation system that may reveal flaws in the system faster than you might like. Also, wiki collaboration can be a tricky balancing act, and if your writing team does not already collaborate and share well, a wiki may cause battles over who can see and edit which content items, and lead to other technology-based secrecy problems that can undermine the spirit of the wiki.

Do not build a wiki if customers do not want one

If your customers have not requested a wiki, there's no need to build one or coerce them into thinking they need one. Instead, "wiki-fy" your documentation set by thinking about other aspects of wikis that you could adopt without actually using a wiki. How about a comment or a rating form on each help topic? Or, set up a method of ensuring that customers or service personnel get credit and perhaps a byline when they make significant contributions? Also realize that your customers may not know all of the features that a wiki could give them, such as notifications on content changes.

Inheriting a large wiki

Some wikis are already in place and in a certain state by the time you are introduced to them. How can you begin work on a large wiki, and what are some areas where you can help by contributing to wiki content?

With the OLPC wiki, an individual contributor, David Farning, went through the wiki and found the following categories. It's an accurate content analysis based on my experience with the wiki. It was impressive and also helped explain my initial bewilderment at how to wrap my arms around the entire wiki.

- Philosophy
- Contributing
- Creating
- Curating
- Projects
    - Project Deliverables
    - Projects In Progress
    - Project Ideas
    - Project Management

After Farning came up with these categories, he asked SJ Klein, director of community content and long-time Wikipedian, if he thought the wiki needed structure. Klein said that a wiki is purposefully without hierarchy (flat), especially for projects, to avoid forcing a parent or sibling sense for projects. He also said, however, that if you have a specific tree hierarchy in mind, feel free to develop the idea in some temporary space. So, when you are working on a large wiki, if you have some good ideas about organization, set them up and then ask for community feedback.

Here are a few more ideas for getting started with a large wiki:

- Although the task might seem similar to starting a large writing project, I have found that wiki editing has some subtle differences because of the collaboration and community "vibe" already present behind the pages. You have to work to figure out that vibe, and then determine your course. New people must get a feel for the community so they can figure out who they might irritate or please by editing a topic and whether they have the confidence to provide correct content.

- Determine your role within the wiki community. It might take a while for you to get to know who's there, what *their* roles are, and where you might best fit in. Introduce yourself with your profile page, following the MySpace wiki pattern.[18]

- Just like a newbie on a writing team, find out if there's some "grunt work" that you can do to get your feet wet and to gain the community's trust.

- Deletions in a wiki can incur some wrath, so they seem risky to do when you are starting out. If you do think some content should be deleted, contact the original author or the big contributors and ask whether you can mark it for deletion. Then, mark it and hope that someone else (a wiki administrator) determines that it should be deleted.

- Start small by tagging or applying templates. Those tasks help you get a feel for the bigger picture.

If you want to coach others who come to your wiki looking to do editing, you might look at the Mozillazine Knowledge Base's guidelines on their About page.[19] For writers who want to come in and "make big changes," their wiki states:

---

[18] http://www.wikipatterns.com/display/wikipatterns/My+Personal+Info
[19] http://kb.mozillazine.org/MozillaZine_Knowledge_Base:About

> If you want to propose changes that affect a
> number of articles, for example a new categor-
> ization system or a generally [sic] reformatting
> of a class of articles, you should make your
> suggestion at [the wiki page] Knowledge Base
> changes. Doing so will give other editors a
> chance to give their opinions on the proposal.
> Even if you don't think your change will be
> implemented, we're still interested in hearing
> your opinions.
> —Mozillazine Knowledge Base[20]

I appreciate this informative and welcoming statement because it
helps contributors understand the way to approach their large wiki
and its community members.

Working in a wiki

Wiki writing versus planned writing

At the simultaneous book sprint hosted by the Google Summer of
Code in 2011, we gathered on the third day to discuss our experiences. One of the participants made a fascinating observation about
the process of writing a book – he actually approached the act of
writing differently during the book sprint week, with more focus
on completion and quality, because he wasn't writing in a wiki. In
a wiki, he expected to be able to write in an outline, freeform, and
someone would come behind him, he assumed, and make the text
better. But knowing that the final deliverable was looming at the
end of the week, he wrote more carefully with an audience in mind
when he wrote in the FLOSS Manuals Booki tool.

---

[20] http://kb.mozillazine.org/Mozillazine_Knowledge_Base:About

This observation captured my attention. Do others feel this same way when they contribute to your project's wiki? In my work, I try to avoid that problem by strictly stating what information goes into the OpenStack wiki, in hopes of also focusing better writing attention in other areas. But what about writing for a linear, printed book, versus writing for a website where every page could be the first page?

As experienced professionals, we know we can improve our content by writing for personas and telling stories. But it is difficult to coach others to do so. I felt that the book sprint experience – compressed and yet with long days of intense collaboration – was a great environment for coaching improved writing in others. It's possible that wiki writing requires coaching on writing even more than other writing methods.

One of the ways we encouraged stricter writing discipline was by encouraging story telling. People were fascinated by the idea that a book can tell stories. Part of uncovering a story to tell involves audience discovery.

At the book sprint, FLOSS Manuals community members like Janet Swisher led the teams through an audience analysis exercise. These were her questions: Who is using your tool? Why do they use your tool? What kinds of things are they trying to do? What can you assume they know? What do they probably not know when they approach your tool?

Example: Open MRS

For these exercises, free agents like myself were assigned to one of the four teams. I went to the Open MRS team, which provides medical record software for developing countries. This was a great chance to learn how open source affects human lives. I learned how important it is to know who is setting up their software and why.

It turns out that the majority of the project's communication is with people who want to build a module for OpenMRS in their specialty domain. There are also implementers who want to use OpenMRS as a data store, turning their paper input forms into a form the registration desk can use when a patient reaches the front of the line at the clinic. They've also recently seen an uptick in requests to integrate with the OpenMRS platform, such as inquiries from Doctors Without Borders.

Answering the "who" question also led to answering the "why" question – they use the tool because they need some extension of clinical functionality, they need a customization for a particular strain of drug-resistant tuberculosis, or possibly just because "their boss told them to." The latter then leads to an investigation of why their boss wants to use OpenMRS.

Mobile data entry, data retrieval, and language support were some reasons why they choose OpenMRS. Some of the user stories that came out were not yet well-articulated, but discussing real people with real work was a huge help in understanding OpenMRS.

I envisioned implementers taking an input form and making an online form, studying the data model while they did so. I heard about patients not making appointments but waiting in line for hours at a registration desk where they would be received for an observation by a medical professional, I wondered about the person accepting shipment of boxes of medicine. These every day tasks capture your attention and hopefully can be translated into a better, more engaging read.

Wikis and developer documentation

A research paper by Barthélémy Dagenais and Martin P. Robillard about open source documentation projects titled, "Creating and evolving developer documentation: understanding the decisions of open source contributors"[51], provides additional insight into real-world wiki use for documentation deliverables. In this exploratory study with 22 writers on open source projects, they manually inspected 19 documents from 10 open source projects. To quote from their introduction:

> We also found that all contributors who originally selected a public wiki to host their documentation eventually moved to a more controlled documentation infrastructure because of the high maintenance costs and the decrease of documentation authoritativeness.

My sense is that wikis have a place in certain communities, but sometimes their simplicity may cause them to be perceived as less powerful than other tools. This is especially the case in open source communities, where only open source wiki engines are likely to be used. Wikis may be the first choice, but they are often replaced as content grows in volume and becomes more valuable and strategic

Enterprise-class wikis like Confluence or Mindtouch, which provide more capability and are becoming less wiki-like with each release, are not mentioned in Dagenais and Robillard's research.

Wiki editing

To quote the introduction to the CommonCraft video instructions for wikis, "wiki websites are easy to use, but hard to describe." I recommend their video, *Wikis in Plain English*,[21] which introduces wikis and how they work in less than four minutes.

Learning about different wikis and how they work should involve hands-on exercises as well as reading and watching videos about wikis. Many wikis offer a sandbox area that lets you try your hand at wiki editing.

---

[21] http://www.youtube.com/watch?v=-dnL00TdmLY

Figure 5.1 shows a troubleshooting page on the One Laptop per Child wiki.[22] It has a table of contents advance organizer feature, a graphic, navigation in the sidebar, and a visible search form. The tabs at the top are prominent features of the wiki, enabling users to discuss, edit, and view a history of the page.

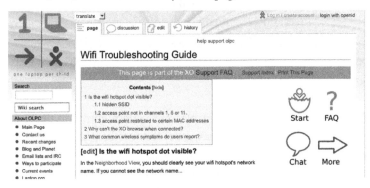

Figure 5.1. Troubleshooting Guide on the OLPC wiki

Even without logging in, a user can click the edit tab in this particular wiki. When you click the edit tab, you see a new interface that shows the editor window. In Figure 5.2, the wikitext is text-based. For example, the three equals signs in a row (===) on each end of a phrase indicate a heading.

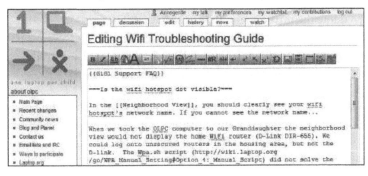

Figure 5.2. Troubleshooting Guide on the OLPC wiki in edit mode

---

[22] http://wiki.laptop.org

Selecting wiki software

The website wikimatrix.org, a comprehensive tool that describes wiki applications and lets you run side-by-side comparisons, uses the categories in Table 5.1 to evaluate wikis:

Table 5.1 – Wikimatrix categories

Feature category	Description
Security/Anti-spam	Does the wiki application provide the ability to suppress spam-like edits or discussion?
	Does it require human-readable forms to be filled out to prevent automatic content fill-ins?
Development/Support	Is the wiki open source or commercial? Does it have third-party extensibility? What support options are available? Does support require payment or licensing or is it handled through the community? How healthy is the company providing the application?
Minor changes	Can a revision be marked as a minor edit so that no change notification is sent?
Change summary	Does the platform offer a page view of summarized changes?
Page history	Does a page history store and show revisions made to the page?
Page revisions	Is there control over how many page revisions are stored?

Feature category	Description
Revision differences	Does the platform offer a view that compares the difference between edits made and whether the difference is for just two versions of a page or for all available revisions?
Page index	Does the platform offer an index listing all the pages on the wiki?
Plugin system and API	Can the platform be extended using add-ons? Is there a well-documented Application Programming Interface (API)?
Links	What is the link syntax? Can links be made to other wiki applications?
Syntax features	Is additional syntax, such as mathematical calculation syntax or custom styles, enabled?
Usability	Is a WYSIWYG (What You See Is What You Get) editor available? Can edits be made on a section within a page? What are the methods for enabling discussing about a page?
Statistics	What statistics/metrics are available? For example, recent changes or wanted pages?
Output	Can the platform output print-ready documents, HTML, mobile formats, etc.? Do HTML-based outputs use CSS styling? Can you export content for import into other systems?

Feature category	Description
Language support	Is Unicode supported? What languages are supported?
Media and files	What attachment file formats are allowed? Can you edit files or launch an editor from within the wiki environment? Is this capability modular?

When evaluating wikis, as with any technology tool, determine your business needs and use cases before you look for specific applications. There are more than 130 wiki engines available for consideration, so you will need to narrow down the list based on your organization's needs. Here are some things to consider:

- What are the primary roles for the wiki: readers, authors, and administrators?

- What is your budget?

- Do you want a wiki that uses a database to serve up the content?

- If you need access to the content for repurposing, versioning, backing up, or translating, how easy is it to get access to the data structures?

- Are you more comfortable with database commands or file copy commands and automation?

- What license is the wiki offered under so you can decide if creating new deliverables from the wiki source and sharing the content is allowed.

- What is the editing syntax?. Some wikis use HTML and some use wikitext for editing. Wikitext is a markup language that usually simplifies the markup further than just HTML, using conventions such as asterisks * for bullets and square brackets

[] to indicate linked text. Will your users tend to spend time in a WYSIWYG editor, or will they be more comfortable in a particular wikitext, and if so, what wikitext do they already know? Do most of your users already know rudimentary HTML markup?

- How about scripting support within the wiki? Some wiki applications enable programmatic solutions within a page, such as creating a table of contents automatically or enabling embedded pages within pages. Are you going to have editors who will want advanced features that scripting offers, such as embedded expanding tables of contents or publishing items from an RSS feed? Scripts may solve some web publishing problems, but for a casual wiki editor, scripted instructions add to the complexity of viewing the wikitext.

- What is the linking syntax? Like HTML help systems, wiki syntax should give you the ability to easily link to other wiki pages and to external sources. Be aware of how difficult or easy it is to link to non-wiki resources.

- Can you set up RSS feeds that enable you or other wiki users to "watch" what pages are being edited? Find out the granularity of changes you can track – is it per article, per page, per category, per section?

- What kinds of search are supported? While setting up search engines within a traditional help system is often difficult, most wiki applications build in easy search mechanisms within the body of the wiki. Find out if you can integrate another search engine with the wiki application or if you must use the one shipped with the wiki.

Managing wiki source content

Single-sourcing – writing once and publishing the content to many different formats – has been a popular documentation method for many years. Many technical writers already write in a single-sourcing environment and wonder if a wiki is just another output. I'd suggest different approaches for wiki as source and wiki as output, depending on the type of content and the community that you expect to contribute to the wiki.

Wiki round tripping is the conversion from a source repository to a wiki and back in order to include edits done in the wiki environment. It may mean processing information files in XML or XHTML, generating output to a wiki in wikitext, collecting the changes on the wiki, and folding them back into the source files – effectively taking the content on a trip around from source to wiki to source. Here are some different ways to manage wiki source content that you might consider:

- **True round tripping:** Edits are allowed in both the wiki and the source repository. This is feasible if you don't mind edits coming through the wiki back into your source. This type of round trip essentially means that the wiki is a source of content and edits, just like any other editing software.

- **Wiki as an output medium:** Every time you publish the source, you publish to a wiki. In this type of wiki creation, the ability to view a revision history is essential for the wiki community to understand what changed.

- **Wiki content published to other media:** The wiki is static and published as is to another format, such as a book or PDF.

You can populate a wiki with content that is written in an easy-to-use authoring system with which your writers are familiar. This method is an excellent way to get an active wiki up and running. However, if you intend for the wiki to be more than just a fancy help system with comments, and you want to allow for community edits, then your wiki has to become the source from that initial publishing point onward.

Another idea in the wiki round tripping arena is using DITA as an in-between authoring-source-and-wiki-storage mechanism. At IBM, they have successfully converted unstructured FrameMaker chapters into DITA topics and then used XSLT to write to MediaWiki for an internal wiki used by their 60,000 services personnel to document complex and specific configurations for optimal performance or for best practices. These special documents are called RedBooks, and the BlueBooks wiki uses RedBooks content to allow experts to collaborate on revisions of existing content. DITA serves as a data conduit, allowing the information to flow into the wiki.

Round tripping wiki content might become necessary if, for example, you cannot adjust to a wiki as your authoring environment, you need to limit authoring to protect content for business reasons, or you are able to gain efficiencies or get a particular output by using an already-familiar authoring tool.

Business reasons for round tripping might include the need to capture wiki edits made by a highly regarded technical person and incorporate the edits into your single-sourced content, the need to generate high-quality print output from a source other than the wiki, and the need to keep the tool "experts" within the non-wiki tool with which they are familiar. Other considerations might include the volume of edits, the need for semantic markup, the amount of "workaround" documentation that might cause contention on an external wiki, and the audience itself.

Single-sourcing and repurposing

Wikis can be sourced from other files or document types, including XML schemas like DITA. If you use single-sourced content to populate a wiki initially, you may also consider round tripping the content. If you're trying to make a help system that looks like a wiki, but you're not taking advantage of the collaborative and crowd-sourcing aspects of a wiki, you just have a help system on the web.

But there are content models that work well with a wiki as the published output from XML source. For example, you can use the wiki as a deliverable and a place to collect comments and post notices, but not allow end-user editing.

You may have another wiki published from XML source where only consultants have editing capabilities. They could contribute to the wiki and edit articles that have been pre-seeded by an information development team. In a wiki for such a focused audience, technical debate about scenarios and implementations may occur, leading to the need to limit edit access. When the editing group is small, the pace and frequency of edits may be small enough to be managed by manually changing your XML source files.

If your wiki becomes the single source, and you output PDFs or wikislices from it, be aware that wiki syntax is, for many, a step backwards to the days of manual markup, and it may not be as quick and efficient as today's authoring tools. For example, you may be limited to a subset of HTML markup, and while you may have access to complex macros, you will need to learn the macro language.

Understanding wiki patterns and wiki structures

Wikipatterns.com is an extremely useful site to help you start wiki writing or maintenance. The companion book, *Wikipatterns*[21], gives insights into wiki adoption and offers valuable case studies. Wiki patterns and other design patterns originate conceptually from architecture patterns.

For example, an entryway pattern is modeled on the entry to a building. Be it arched tile, cobblestone, or a simple wooden gate with a dirt floor, it offers a solution to the question, "How do I enter this structure?" For a wiki, a comparison to an entryway pattern might be the welcoming pattern, where a current wiki contributor leaves a note on a new wiki contributor's "talk" page or sends a welcoming email after the contributor's first edit.

Wiki patterns help you understand the people patterns and adoption patterns for the wiki. There are also "anti-patterns." One adoption anti-pattern is "Too much structure."[25] When you overload a wiki with strict guidelines or lots of empty pages, trying to get people to contribute, you may notice that the wiki does not grow organically like you hoped. In *Wikipatterns*[21], Stewart Mader suggests:

> A wiki will evolve into the optimal organization of information as people use it, and it's better to adjust the structure based on the content, instead of the other way around.

If a wiki is meant to be a content deliverable, strict hierarchy may work just fine to help readers find content. But a wiki that you expect

---

[25] http://wikipatterns.com/display/wikipatterns/Too+much+structure

to have contributions to should have the flexibility to grow as the community sees fit.

Alternatives when a wiki is not the right match

If you determine that your business objectives do not warrant a conversation or community-based documentation, such as a wiki or customer-focused blog, you might consider different ways of implementing changes to your current documentation output or methods.

Online help possibilities

If you are evaluating tools for online help, you may want to seek out online help tools that integrate comments. An important first step may be to get your user assistance on the Internet. If your online help is not searchable because it is not on the Internet, consider ways to put it online.

Consider a feedback server or Javascript solution that offers tools to enable comments and other social web integrations. You may want to implement a rating system for each help topic or a comment form for each topic. Wiki applications already provide these features, but you may be able to integrate a rating system using existing web content tools.

If you decide to share the contents of your online help by licensing it with a Creative Commons license, you could embed the CC image at the bottom of each page and change the licensing information page in your online help.

Table 5.2 offers some questions to ask about your online help system to look for possibilities for integrating conversations or communities.

Table 5.2 – Online help categories

Category	Description
Online availability	Can search engines find your content?
Ratings	Can readers rate the usefulness or accuracy of a help topic?
Comments	Can readers comment on pages or paragraphs?
Tagging	Can readers tag the content, or can authors tab the content? Can you display a tag cloud?
Collaborative authoring	Can readers edit help topics?
Identity providing	If comments or collaborative authoring are enabled, can you check a collaborator's identity?
Metrics	Can you track page counts and other community membership measurements.
Subscribing to updates	Can readers subscribe to RSS feeds to be notified when the help is updated? Can they get email notifications when updates occur?
Publishing timing	Can you automatically time or delay publishing of help contents?
Integration and sharing with social networks	Can you share help content with other social networks such as Facebook, delicious.com, or Twitter?

Talking with writers of wikis

You can learn from writers who document their products with wikis. This interview is with Dee Elling, who maintains a wiki about Code-Gear for Embarcadero. While researching my 2007 STC Intercom article about wikis and technical documentation ("The 'Quick Web' for Technical Documentation"[53]), I interviewed Dee via phone and email because she left a helpful comment on my talk.bmc.com blog entry about a DITA and wiki combo.

Dee is the manager of the documentation group at CodeGear, and she blogs at http://blogs.codegear.com/deeelling/.

> Interview about wikis for technical documentation
> with Dee Elling of Embarcadero, July 26, 2007
>
> 1. **What are some of the factors for selecting a wiki
>   software package?**
>
>   I've encountered hesitation from some writers
>   about using a markup interface. Many writers pre-
>   ferred a Word-like GUI interface, such as Confluence
>   provides. Another consideration is cost, since there
>   is not always a budget for new systems; at CodeGear
>   we use MediaWiki. Primarily we manage internal
>   information such as schedules and doc plans; lately
>   we are collaborating with engineers to write FAQs
>   and release notes.
>
>   At my previous employer, one engineering team
>   was writing the documentation themselves on the
>   wiki (using outlines provided by the writer), and the
>   writer cleaned it up and converted it to PDF for
>   distribution with the product. That is a great use
>   case which I believe could seed the adoption of
>   wikis into the documentation process, especially in
>   companies where there are limited doc resources.

At CodeGear I can post copyrighted material to our Developer Network. The Developer Network technology allows comments on postings, which is not the same as wiki but a good start. Since joining, I have started to post traditional doc content as "articles" there. I've already fixed a few doc issues due to rapid customer feedback! We are also working on a design to make the website interface more wiki-like.

2.   **How do you get legal approval for such an open-edit site?**

At my previous company I never got to the stage of implementing a public wiki. However I had many discussions about the legal aspects related to the product documentation. The legal aspects seem complex, but lawyers can write new terms for new situations.

At CodeGear the issue will involve intellectual property but the user base is so active on the internet that there are few "secrets." More important will be the issue of releasing information too soon or otherwise getting in trouble with SOX compliance. (That makes my head spin!)

3.   **What are the considerations when choosing where the wiki is hosted?**

Cost and reliability are factors, but most important is buy in from the IT department, who would likely manage the hosting.

4.   **Which types of products are best targeted for a wiki?**

Complex software products are a good example. There is so much flexibility in software, and product documentation cannot cover every use case. The wiki lets customers add content that is relevant to their own use cases, and that will benefit others.

5. **How can you encourage your users to contribute?**

Keeping up a dialog with the customers is helpful. If you respond to them, a dialog develops, and they are more likely to contribute again.

6. **What are some of the success factors for the wikis that contain technical documentation?**

Does it result in more positive feedback from customers? Do customers help each other and contribute to strengthening the installed base? Does it increase product visibility and mindshare in the market? Is it perceived as a strategic advantage over competitors? Does it cut down on tech support phone calls?

7. **What traps should be avoided?**

The trap of not responding or not paying attention. Writers must diligently track the "living documents" they have created, and they must truly collaborate. If customers contribute and their contributions go unrecognized, they may think that the company is not fully supporting them.

Public wiki documentation must be actively managed and that takes more writer effort than in the past, when documents were forgotten as soon as they went to manufacturing. Contrary to the fear that with wikis you don't need writers anymore, I believe that with wikis the role of the writer will grow.

Here is my second interview with Dee Elling, from 2011.

Interview with Dee Elling, 2011

To give an update of her perspective, Dee agreed to a second interview with the new edition of this book. At Embarcadero, Dee spent two years with help systems delivered through complex wiki systems using Mediawiki. You can see the site at docwiki.embarcadero.com.

1. **How are you aligned in the company? Support, marketing, training?**

   The doc team is aligned with the software development team, and works very closely with the localization team. The product is developer-focused so most customer interaction is directly with the dev team.

2. **What is your job title?**

   Senior Manager, Information Development

3. **What types of content do you deliver?**

   Topics about programming concepts and using the GUI development tools, plus detailed language library reference topics. The language reference information is partially generated from the language compiler and automatically merged with manual content from the writers.

4. **How is your content licensed and how was the license selected?**

   It is a customized copyright, no special license. The higher-ups did not want to give up the copyright, and wanted to restrict mirroring or other wholesale copying. This would have been the case with a general website, too. All contributors implicitly agree that Embarcadero holds the copyright.[28]

5. **How many collaborators do you work with regularly?**

About 12 offshore technical writers and translators, 1 local writer, 5 or so local internal contributors from the Development team, and 8 or so active experts from the community. Many more than that view the content.

We get the most feedback from Beta customers; that feedback/conversation happens mostly via forums and email, and some directly on the wiki.

6. **Do you hold sprints and if so, how many people are active in sprints?**

Sprints are 3-4 weeks, but due to internal issues, the process is very poor for documentation. It's not uncommon for the scrum masters to allow documentation to miss the sprint deadlines, and then ignore it. Most documentation comes along after the product release – that's another reason to use a wiki instead of a more static system.

7. **How many page views do your pages get?**

about 12K/month, 400/day across 3 release versions in 4 languages.

8. **What is your tool chain and how is it created? Internally, with outsourcing, through the community?**

The internal webmaster/wiki developer manages a Mediawiki farm. He handles the conversion of the language compiler information to wiki format. He

---

[28] http://docwiki.embarcadero.com/RADStudio/en/RAD_Studio_XE2:General_disclaimer

also handles the customized Python/JSON/PHP system that generates the Help2 that is shipped with the product GUI.

9. **Anything else you want to share about your particular situation?**

The writers are very happy using the wiki, except when the network is slow. Their feelings about writing "in the open" are mixed; some of them are very assertive and just write away; others are more timid and write offline before putting it on the wiki.

I have frustration with people who won't share their knowledge with those who know less. Some people still have old-school habits of hoarding what they know, as if their jobs will be more secure if no one else understands. Or maybe its ego. Either way I just don't get it.

I used to get frustrated with people who will use the wiki to read but then send emails or write forum posts instead of just commenting on the wiki. Now I understand that email and forums give people a sense that someone "is on the other side" and will get the message; that sense of personal contact is more diffused on a wiki. Now I see that the wiki gives readers up-to-date access to what the writer is in the middle of doing, and that emails and forums are an equal if not better way to have related conversations with a larger group of people. Also, people **really like** personal responses to their emails. Documentation@embarcadero.com is a very busy user! Putting that email on every single page helps generate a lot of conversations.

Wiki examples

This section contains examples of some of the highest profile wikis in use today for technical documentation, targeting endusers of a product or service, or assisting developers in their tasks. Although only a small number of technical documentation wikis are publicly available, I believe that many more wikis are used internally.

Adobe Labs

The Adobe Labs wiki[29] (Figure 5.3) targets developers using Adobe technology and is one of the most frequently used wikis according to Alexa rank.[30] This example shows how a community can build up around a wiki – developers helping other developers accomplish tasks. This type of wiki can build communities of practice more quickly than was possible before the introduction of wikis.

Figure 5.3. Adobe Labs wiki

Apache

Apache is a collection of open source tools that run the majority of web servers on the Internet. Apache is one of the best-known and successful open source software projects, and it is a natural choice for its documentation to be available in a wiki (see Figure 5.4).

---

[29] http://labs.adobe.com/wiki
[30] http://alexa.com/site/ds/top_sites?ts_mode=global&lang=none&page=1

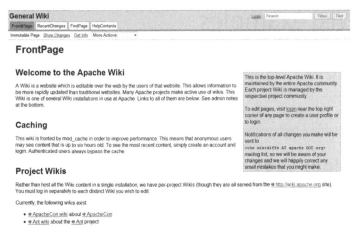

Figure 5.4. Apache wiki[31]

Microsoft Developer Network (MSDN)

Microsoft's Developer Network (MSDN),[32] Figure 5.5, has been watched closely as one of the first forays into community-contributed content, but as it has evolved over the last few years, it barely qualifies as a wiki by most wiki standards (see Mike Gunderloy's article, "The wiki-fication of MSDN"[55], for more).

Rather than allowing any article to be edited, MSDN allows only certain highly structured contributions, such as code examples. The system strongly types the information to be entered in a web form. Also, the MSDN wiki's top contributors are Microsoft employees rather than outsiders to the company. Still, many contributions are made to the wiki-like system, and it is important to watch and learn from large-scale deployments of user-generated content.

---

[31] http://wiki.apache.org
[32] http://msdn2.microsoft.com/en-us/library/default.aspx

Figure 5.5. Microsoft Developer Network

OpenDS on java.net

The OpenDS wiki[33] (see Figure 5.6) was created because of the open source nature of the OpenDS product, and the wiki is the only source of documentation for the product. Three writers are paid to maintain it, just as any technical writer is paid to create and maintain documentation in any other tool.

The writer Ragan Haggard has presented a few times[34] on his experience with wikis for documentation, and he likes the easy access and simple interface that the wiki, which uses the MediaWiki application, offers. However, community participation has not been growing.

---

[33] http://www.opends.org/wiki

[34] https://www.opends.org/wiki/page/RHaggard

This type of response has been common for wiki roll-outs related to product documentation – apathy is more prevalent than any type of vandalism or malice.

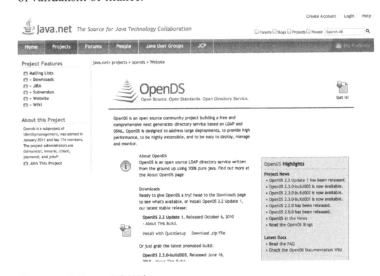

Figure 5.6. OpenDS Wiki

Cisco DocWiki

After Cisco first rolled out a customer support wiki and experienced some success with it, the Cisco documentation team was asked to create a wiki for end-user documentation (see Figure 5.7). Cisco conducted a careful pilot project, allowing internal-only access, before releasing the wiki for external use.

Figure 5.7. Cisco DocWiki[35]

Wiki wrap up

While organizing the content and picking the right tool are important, there are many other implications when using a wiki for information delivery. Understand that the 90-9-1 rule means you will have
a small club of one-percenters, and that one-percenter may be you.
The community built around the wiki is as important or more important than the tool used to build the wiki. Because wikis are not
like traditional documentation, strict style guidance and strict hier-
archies may not work for the community that embraces the wiki.

---

[35] http://docwiki.cisco.com

6

Finding Your Voice

Writing gives you the illusion of control,
and then you realize it's just an illusion,
that people are going to bring their own stuff into it.

—David Sedaris

Standing out in the attention-hungry blogosphere means you must write better than most of the crowd. Not only must you *write* better than the crowd, you must also *communicate* better. This chapter describes best practices for writing when using social applications and communicating with people who prefer to obtain their technical information through the social web.

As Muhammad Saleem wrote on Copyblogger, when "writing specifically with the social media audience in mind, understand the mentality and what they're looking to get from Digg, Netscape, Reddit, StumbleUpon, and so on, and appeal to that desire."[78]

Style guidelines

For the most part, "conversation" is a less formal style of writing than technical writers are used to. You may need to write a separate style guide for blog, screencast, and tutorial deliverables, or create a subsection of your current corporate style guide to discuss the special rules that apply when writing for social web media and audiences.

An excellent style guide for online blog articles comes from A List Apart,[1] a collection of essays and articles for web developers who are experts in CSS (Cascading Style Sheets). Informality is not the first goal of the style guide, but rather clarity, audience consideration, and links to definitions for those readers who are well-versed in code but may not be familiar with the vocabulary of an industry expert in visual design or any specialty the blog's author discusses.

A minimalist, clean style in your writing is best for reading on the web. Brevity may seem curt when speaking to someone, but if the writing is more clear when brief, don't think you need to embellish the writing with lots of small talk or jokes.

A sense of humor can come through as part of your personality, as long as the humor matches your audience. A classy, clever sense of humor will be welcomed by the types of readers who read blogs for information, and you can tailor that sense of humor (and adjust your definitions of classy and clever).

---

[1] http://www.alistapart.com/contribute/styleguide/

You do not need to use the formal trademark and registered trademark symbols on first use when writing for online content like a blog. It's more difficult to determine where the "first use" occurred anyway. But do be respectful of brand names and company names.

Many times people assume that writing for the social web automatically calls for a more informal style in writing. Some might even predict that our communications in the future will include "text speak," the shortcuts and slang in the English language often used on mobile devices where it's faster to type short phrases and use numbers or single letters with phonetic pronunciation. Tony Self ponders this issue in "What if Readers Can't Read?" [79]. I think that professional writers should stand above that crowd and instead adopt a strict adherence to correct grammar and traditional spellings; however, a casual, speaking-like style in your writing is perfectly acceptable and likely preferred in most communications.

In *Letting Go of the Words: Writing Web Content that Works*[27], Ginny Redish talks about a casual but concise writing style. Microsoft's most recent style guide requires a specific tone and voice:

> Use the Microsoft Windows tone to inspire confidence by communicating to users on a personal level by being accurate, encouraging, insightful, objective, and user focused.... Don't use a distracting, condescending, or arrogant tone. Avoid the extremes of the "machine" voice (where the speaker is removed from the language) and the "sales rep" voice (where the writing tries to sell us something, to cajole us, to cheer us up, to gloss over everything as "simple.")
>
> —Microsoft Style Guide[2]

---

[2] http://msdn.microsoft.com/en-us/library/aa974175.aspx

These are excellent guidelines to follow when establishing your own balance between casual and formal.

Open and honest

The social media audience expects open communication and honesty. Much like the general web-reading audience, this audience has little time or patience for messages that are not directly related to their task at hand. Plus, any exaggeration or vague references can be commented on or edited immediately, so be as honest and open as you can in your communication.

At BMC Software, I was instructed by the talk.bmc.com blog team to make my blog sound like I was standing beside the customers, solving problems like theirs. I was encouraged to be honest and open about weaknesses and gaps in my own knowledge about Business Service Management, one of our core corporate competencies.

In a blog, it does not help the reader if you write about things you don't know. However, you can talk about your own learning journey. Some readers may be at the same level of knowledge as you were when you started out. You can only be a "newbie" for a while, so it's valuable to capture that experience and learning path on a blog.

I was also guided to the *Cluetrain Manifesto*[18] and encouraged to advocate its principles through blogging for BMC Software. Here is a powerful quote from the manifesto that strongly states the need for companies and their employees to be direct in their communications with each other:

> ... learning to speak in a human voice is not some trick, nor will corporations convince us they are human with lip service about "listening to customers." They will only sound human when they empower real human beings to speak on their behalf.... employees are getting

> hyperlinked even as markets are. Companies
> need to listen carefully to both. Mostly, they
> need to get out of the way so intranetworked
> employees can converse directly with internet-
> worked markets.
>
> —*Cluetrain Manifesto*[18]

So, does openness and honesty also require, or dissolve into, negativity? I don't believe so. You can be honest about a problem without complaining about it or casting it in a negative light. You can strike a balance, though, so that you do not sound overly enthusiastic or plainly ignorant about certain problems. You cannot, however, be open to the point of violating any policy agreement that you've made with your company.

What about transparency? What does that mean exactly? Transparency in written communications usually means that nothing crucial is hidden from view, and it also usually prescribes a straightforward, no-nonsense, sensible communication style. Secrecy is not tolerated or expected from a writer using social media tools or the Internet in general. Know your company policies for trade secrets and other private information, and know when it is appropriate to talk about products still in development. The trigger for when you can write about something will vary depending on your company's policies and its business goals related to using social media for messaging.

Make sure you understand your company's policies regarding trade secrets and other proprietary information. In most companies, you will get in serious trouble if you reveal proprietary information.

Transparent, clear, and authentic writing is a powerful tool for convincing others of your point of view, but it is also important for building trust with your audience. This trust and belief in the accuracy of your documentation is crucial for creating successful relationships with readers through any social media communication.

Personal and professional

When writing for social media, you have to present yourself as a real person. Writing in anonymity won't work in most social media situations. Readers are interested in the person behind the writing, who you work for, why you started writing, and how you came to know the information that you know. Your identity and reputation precede all that you say online because it is generally easy to gather enough information to form an opinion of you. Typically you link from your front page to a page that describes you and your interests and perhaps the reasons why you write a blog.

People seeking information on social media outlets are naturally drawn to seek a relationship with the person providing the information. They want to get to know you. Give them some information to put a face with a name, including a professional picture that reflects your style.

Recently there has been much importance placed on building a personal brand. Although your goal might be to merely enable conversations instead of trying to stand out at your company, be aware that your writing represents you as an individual as much or even more than it represents your company.

Although you should insert your personality into your social media communications, especially for blogs, forums, and presence content such as microblogs, you need to strike a balance. You must consider your own privacy as well as that of your family members, your co-workers, and anyone that might be part of or affected by a personal story you share.

But also be aware of the discomfort that some people may have with expressions of political views, sexual preferences, and religion. What seems innocent and harmless to you may be a red flag to someone from a different culture or age group. Use good judgment and as always, know your audience.

Telling a story

It's important to share personal stories, especially if you are passionate about something in your company or your work life. If a story effectively relays that passion in a "sticky" way, the story remains with the reader. For example, I shared a personal story about getting an eye injury from a piñata bat striking my eye at a children's birthday party.[4] I connected the experience to my chosen profession of providing web-based information and linked the post to accessibility guidelines in place for low vision people who want to get information from websites.

Because it was such an unusual incident, and because it related to me and my family, including my children, some might consider it risky to relate such a story on a blog. But if the story helps someone remember to enter ALT tags on images so a screen reader can help someone with low vision understand a web page, and if the story shares in a professional way without too much detail, I believe it has a place on a blog.

Snappy titles

To encourage readers to go beyond skimming headlines and click through to your content, you should borrow some advice from copywriting and journalism schools. Although titles should be accurate and convey the correct information, you also have some freedom to create titles that are more enticing.

Copyblogger author Brian Clark talks about the Swipe File, a collection of headline templates that capture the readers' attention in two separate blog posts.[5] He suggests going beyond the common "How

---

[4] http://justwriteclick.com/2009/01/28/arrrrrrr-mateys/

[5] http://www.copyblogger.com/10-sure-fire-headline-formulas-that-work/, http://www.copyblogger.com/headline-swipe-file/

to" and "List of" titles to formulas for headline writing. While you don't need to sprinkle copywriting throughout your technical documentation, you may want to study these types of headlines and determine if they might work for your social web writing efforts. Writing an enticing, attention-grabbing headline takes practice but may make a difference in number of visitors.

Publishing strategies

You are a publisher each time your content goes on the web. You need to shift your thinking away from print writing towards web writing, and you need to think about publishing strategies before you write a single paragraph. When I first started blogging, writing two posts a week was my intention. This strategy meant I also needed to determine the spacing and timing for those two posts. This section discusses some of the considerations when blogging and microblogging for publishing.

Timing

Daily newspapers are typically available in the morning, but the constant availability of today's news makes the idea of waiting until breakfast time to get your news ludicrous. Because the social web encompasses the world, people in different time zones wake up at different hours of your daytime and scan the social web for information and updates.

For Twitter posts, you can set the time that a Twitter update is published using a site like SocialOomph.com. You can publish when you believe your readers are first reading their information or you can time the publishing to catch them later in the afternoon or evening, but also realize that the earth is round and you may need to post multiple times to set your message timing around the globe.

On her WordPress blog, Lorelle VanFossen has an excellent post titled "When is the Best Time and Day to Publish a Blog Post?"[87]. She writes about analyzing when your readers tend to read posts and when they tend to comment on posts. Use analytics on your website, blog, or wiki to find out when people read the content, and then time the publishing of your content so that it matches when your readers are seeking updates or searching for answers.

Some bloggers have even set up specific "open comment" times, such as on Tuesday evenings. Readers know they can go to the blog site at that time and post comments and get immediate responses from other readers and the blogger.

Search engine optimization

In the economy of web content, you are paid in links to your content (which indicate your reputation) and in attention (which is indicated by the number of page views your site receives). Both of these "currencies" can be optimized and are part of a search engine optimization (SEO) strategy.

Most blogs already follow the basic rules for optimization – content is updated often and tagged with keywords; natural, readable URLs are used; importance is placed on the first paragraph, heading, and title text; and the content is text-heavy so that it is not hidden behind Flash introductions or video.

However, because buzz generation is not the main business goal for most documentation, search engine optimization is probably not important enough to make you switch your online help engine to a blog engine. But you should evaluate the HTML that is generated from your online help tool to ensure that it can be optimized so that it can be found.

Another key to optimization is to ensure that your domain name is reserved for several years, not just one year at a time. To search engines, this indicates that your content will be available for a while and is valuable enough to keep archived. You should also ensure that your content is registered with search engines. There are many search engine submission sites, such as addme.com.

SEO techniques are constantly changing. Google has a Help Center for web masters[8] with information about how Google works, and you may want to look for similar information from other search engines.

Submitting to social bookmarking sites

You should make your content sharable – blog-able, tweet-able, and tag-able – because you enable others to promote the content for you. I may owe the entire genesis of the writing of this book to a single bookmarking event by a co-worker of mine at BMC Software.

On my blog at talk.bmc.com, I interviewed a fellow technical writer about working on a wiki for a cell phone manufacturer. Will Hurley submitted it to digg.com because he found it interesting. About a month later I had an inquiry from the editor of *Intercom* magazine, the magazine for the members of the Society for Technical Communication, asking if I could write an article about wikis in technical publications. I obliged, and it was the start of a great research and experimental phase, thanks to the social tagging of that article on digg.com.

You might find it useful as a publishing strategy to tag your content for social bookmarking sites. Other users may follow suit and your keywords will have multiple submissions that will help them stand out in the crowd.

---

[8] http://google.com/support/webmasters/

Finding your audio voice

You may also find that text alone will not help your message reach as far. When I was at BMC Software, the coordinators of the talk.bmc.com site decided to do two separate podcasts with the two most popular bloggers on the site: Steve Carl and me. It was my first podcast and I was extremely nervous but very excited about learning a new way of communicating. I chose to go to a home-based recording studio in south Austin, with padded walls and excellent microphones. We had a fun time, and I truly felt comfortable with my interviewer Ynema Mangum.

Four years later I learned that podcast was the one that fellow technical writer and blogger Tom Johnson heard first, before he started to read my blog at talk.bmc.com. Since Tom was interested in podcasting, and because podcasts were a more "unusual" form of communication for technical writers, the podcast came to his attention first. Be on the lookout for similar technologies or publishing venues that might help your other deliverables gain attention.

Link to your work

While I was blogging for BMC Software, I was also writing white papers that described complex cross-product solutions to particular business problems. I liked to write a blog entry talking about the white paper and the problems it described and then link to the white paper download site once it was available on bmc.com. On a corporate blog, this type of linking helps to show that your first priority is your work, and that your work is relevant to the types of problems your readers face.

Idea generation

Three basic steps can get you started with social technologies and documentation:

1. Listen
2. Join
3. Provide a platform

Listen to the conversation

This section gives ideas for reading, listening to, and monitoring online discussions about your product, your company, and the people who use your product or service.

■ To find social media content related to your company, search for your product names and company name on sites where Web 2.0 conversations occur. Use a Yahoo Pipes Social Media setup to monitor the conversations about your product in the blog world, on Twitter, and so on. In her blog article, "Keeping up with the social media fire hose"[63], Jackie Huba describes a pre-configured Yahoo Pipe you can use to find out more about what is being said about your company or product.

■ Sign up for notifications on customer support forums that are relevant to your area of expertise on your product. Monitor the forum and step in to help whenever you can. Look for conversations on email lists beyond your company's customer support forums, such as a Yahoo Group dedicated to your product. Point to your online help systems when appropriate.

■ Use Technorati and BlogPulse to monitor keywords that are pertinent to your company and products as well as your competitors. Use Search Engine Optimization (SEO) keyword tools like SEMRush and the Google AdWords Keyword tool to study and research relevant keywords.

- Constantly read and monitor comments on articles in wikis and support forums related to your products. Often the true value to the customer is in the discussion on an article or post about strategy or best practices, and not in the item itself. Questions are asked and answered, and nuance is explored and interpreted through conversation.

Join the conversation

Once you are adept at listening and understanding the conversation and community building going on, you can join in by talking to others or providing information on a community site.

- Start a Flickr photo set that contains photos of your product in use. If your product is software, perhaps use screen captures. Be sure to offer navigation hints for how to open a dialog box or set up a particular view. For hardware, photos are especially helpful for providing how-to information, such as opening the case on a computer, inserting an SD card, changing a battery, or loading an ink cartridge.

- Post screencasts on YouTube or a higher-resolution video site like Vimeo or Viddler. Screencasts are video tutorials that take video of your screen while a voice over explains the task being accomplished. If operating your hardware product has a tricky physical task, you might post a video of someone performing that task.

- Future-proof a helpful set of URLs by creating delicious.com tag sets that describe how-to information about your product. Even if you add additional tagged URLs, delicious.com groups them by tag so people get the latest set of URLs along with your notes about the links.

- Start a blog at your company, or write entries as a guest or a regular contributor for an existing blog at your company.

- Offer comments on a customer blog when the blogger writes an entry that is relevant to your product. When a customer finds your blog, respond to comments quickly with an answer or simple acknowledgement.

- Contact a customer and ask if you could interview him or her for a podcast. You can post the podcast on your blog as an MP3 file or host it on one of the podcast websites.

- Ask a customer to write a Wikipedia entry about your product's technical concepts or submit one yourself. But be aware that Wikipedia does not accept articles for commercial products because articles must pass its notability test. Before submitting to Wikipedia on behalf of your company, refer to Wikipedia's list of policies.[15]

For example, ASI's solution architect, Jay McCormack, submitted an article defining "friendraising." Friendraising is the practice of online event-based fund raising using social networking and viral marketing techniques. In less than an hour after his submission, the Wikipedia community had altered the article presenting a much more cynical view of the term, plus marked it for deletion identifying it as a neologism without reliable sources and verifiability.[16] The article has survived for more than a year, though, and is no longer marked for deletion.

- Set up a Facebook group or a LinkedIn group after determining where your users are most likely to join in. Building this type of group on an existing platform helps you connect customers to each other with minimal effort.

---

[15] http://en.wikipedia.org/wiki/Wikipedia:Policies_and_guidelines
[16] http://en.wikipedia.org/w/index.php?title=Friendraising&action=history

Provide a platform for conversation

Even though only about 1% of the people in an online community will be contributors, you may find that it makes sense to provide a platform especially for them and other community members who will benefit from their contributions. Here are some ideas:

- Provide a user group email list using Yahoo Groups, Google Groups, or any of the multiple choices for email lists, or ask a known user community member to do so. Realize, though, that mailing lists can make group work more difficult.

  As Clay Shirky eloquently explains in "Group as User: Flaming and the Design of Social Software"[80], mailing lists can have the most difficulty with signal-to-noise ratio because members have difficulty obeying rules of conduct when there is so much communal attention (both positive and negative) available. When the list is large, the problem is more difficult to manage.

- Provide a commenting platform for customer using your online help system or a user assistance wiki.

- Provide a full-fledged wiki with editing enabled to give customers a chance to write or edit pre-seeded articles.

- Offer a video and multimedia sharing platform like Sun Microsystems does at http://sunfeedroom.sun.com.

- Give users a blogging platform by either building a planet blog (a collection of blog entries from different bloggers using aggregated feeds) or hosting customer blog entries on a website.

- Build a social networking community using hosted software like Ning.com (an online platform for people to create their own social websites and social networks) to allow customers to communicate with each other and your company's employees.

Living and working with conversation and community

Stewart Mader's consulting brand is called "Future Changes," and I like this turn of phrase. It can either describe changes in the future or offer a mere statement of fact that the future changes. The future of education, business, and technical communication has changed, is changing, and will change. Right now it seems like the number of social media tools is on a constant upward trend, but social methods for documentation may take a place in an overall content strategy as another publishing mechanism. It will be exciting to see how it evolves over time.

In the workplace, I have always surrounded myself with people who are smart, professional, and have an eye on the future. Collaboration, innovation, and creativity are high priorities in my work environment. At BMC Software I sought out new areas for communication when my job description charged me with doing so. I was fortunate to find people who were breaking new ground at talk.bmc.com.

As an indicator of their forward-thinking career paths, some of the people experimenting with blogs and podcasts in 2005 are now working on cloud computing solutions. Others are inventing remote video interviewing techniques and creating media channels where there were none previously.

People ask me, where should we start with these conversational and connective methods? In 2005, I started with blogging behind the firewall. But in 2009, writing Twitter posts with protected updates might have been a good starting point. In 2012, Twitter is still taking the world by storm, but Google+ may point the way to your starting line.

People also want to know, where should my team start conversations? Or where should we focus our time if we do start? In this book, I talk about phases: Listen, Participate, Share, and Build a platform. I think you should start with listening and monitoring what's already being said. Next, start commenting on blogs, blogging, or even microblogging yourself.

A small step towards blogging is to blog on an internal site, behind your firewall, to limit your audience if that makes you more comfortable. I'd recommend trying out tools that are already installed that you don't have to maintain and install yourself.

For example, I started my blog, JustWriteClick.com, on wordpress.com and paid $10 a year to map my domain name. When I knew WordPress was a good fit for me and my blogging and site needs, I found a hosting provider and installed WordPress on the server myself. If blogging seems too big a first step, start with reading and commenting on blogs or simply add a comment mechanism to your documents site.

Sharing content is the next step, and you can learn so much by sharing content and talking with others about it. Full time, continuous engagement with a user community while sharing content must be preceded by real relationship building with users. These steps require that you know your company's social media strategy, which may be a content strategy with a social media delivery mechanism.

The final step is providing a platform for users. These steps take time but you will learn valuable lessons along the way and hopefully avoid any stumbling or disastrous results. It's okay to fail, though. You learn new lessons with each attempt and approach. And you meet people, have ongoing conversations, and help build online communities along the way.

Content Strategy for Community Documentation

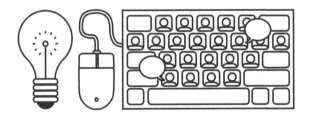

*But without ambition, desire, and focus, social media
is a recipe for chaos. Through all of the distractions and fatigue,
we must continually renew our focus to bring important goals to life
based on our actions and words in each social network.*

—Brian Solis

Content strategy as a career path, business strategy, and discipline,
has taken huge leaps and strides forward in the last few years. Ann
Rockley published the first edition of *Managing Enterprise Content:
A Unified Content Strategy*[29] ten years ago, but the practitioners

did not call themselves content strategists at that time, and even now just a few practitioners use the title.

There are small islands of content strategy teams at unlikely non-software companies such as Wells Fargo, but the software giant IBM has also adopted the title in its career paths. At Rackspace we have content strategists on support teams, creating customer-facing technical content but we also have technical writers in the trenches of the development organization.

Content strategy themes continue to grow due to evangelism by information architects who realized they were actually doing the work of content strategy. Examples include Karen McGrane, high-energy personalities like Kristina Halvorson, and early proponents of the field like Ann Rockley, who first realized the value that content gives to business and wrote a comprehensive guide to content strategy for print and online deliverables (*Managing Enterprise Content: A Unified Content Strategy*[29], now in its second edition).

Just as technical writing has evolved away from massive printed manuals to lightweight online deliverables, so have all types of writing evolved towards producing the deliverables our audiences expect to see, with production orchestrated by content strategists. People bring certain expectations to web content such as sharing, perhaps liking, and certainly being able to comment on a page or tag it for retrieval later. This evolution means that today content strategy is being defined more deliberately than ever before.

If I were in college today studying technical communication, I would want some practical training in content strategy. If a blogger like Heather Armstrong of dooce.com can make $40,000 a month with blog entries, photos, and building a community, then you know the money is available to hire talented web writers who can help you reach these goals through content strategy.

Content Strategy for the Web[12], a slim, tightly-organized book by Kristina Halvorson, prescribes a methodology that involves creation, distribution, and governance of web content through planning. This book is also available in a second edition. Halvorson and her colleagues contend that a company's websites shouldn't be treated like the interior decorations in a lobby – giving them a facelift with new carpet and furniture every couple of years and expecting your business will continue to function as usual won't work.

Websites that truly serve the needs of a customer are much more complex than the user experience of entering a building's lobby. By creating personas, doing task analysis, and creating content that serves the needs of those personas and creates a wonderful user experience, businesses can tie their success to the success of their website.

While her book mentions user-generated content, she gives a nod to the difficulty of using it, saying,

> This is a fairly complicated, surprisingly resource-intensive approach to sourcing content. If you build a user-generated content forum, it doesn't necessarily mean that they will come. And if they do come, it doesn't mean they'll stick around. Engagement tactics are key, as are resources that will moderate and respond to content and comments.
> —*Content Strategy for the Web*[12]

These engagement tactics are at the heart of content strategy for communities. According to the Community Roundtable Report from 2010, "The 2010 State of Community Management: Best Practices from Community Practitioners"[50], community managers have found that you can plan and budget for different balances between user-generated and professionally-created content to fulfill different goals for your site. When searching for solutions that are crowdsourced, always consider the business goals and timing for getting there. It may be best to accept imperfection in the deliverables in order to get content out there quickly.

Another community content tip drawn from the Community Roundtable Report is related to single sourcing. Rather than just single sourcing text, you can re-purpose community content by creating a podcast interview with an author. The source is still the expert or story teller, but the output can go beyond just text. They have found that relationships are not built with text alone. Storytelling through pictures and videos helps bring engagement to a community-based site.

Listen and monitor first

To me, content strategy for community documentation involves listening first, then determining whether a community strategy makes sense not only for your business goals but also for your audience. Rather than thinking about all the posting and publishing methods for your community, think about how you will gather feedback and discussion points with community members. A listen-first mentality means you set up Google Alerts for keyword searches on websites, news sites, and blogs. Seeking blog posts that use the most relevant keywords can help you connect with bloggers. Listen first also means you are constantly looking for authors and content that your community interacts with, welcomes, and reads often.

Find the business goals

There are many ways that good technical content can help a business. For example, imagine if you could track a marketing campaign to get 100 more subscribers to a service to measure your content's effectiveness. If you find a way to add 100 new subscribers quarterly, you are contributing to the bottom line of the company and its growth.

Maybe user assistance and help text is so tightly integrated that customers rave about the seamless experience and come back for more because they know they can get in, get what they need, and get out quickly.

Speaking of efficiency, how about measuring a decrease in support response times? Or possibly measuring how effective your users are at providing peer-to-peer (unpaid) support because of the doc comments platform or wiki platform your team created? Refer to Chapter 8, *Analyzing and Measuring Web Techniques*, to find more ways to correlate content effectiveness with business goals.

I find the analysis portion of content strategy very exciting – we are living in a time where the content grades itself if you use web analytics. Does your content get an A+, at the top of its class, or a C-, barely passing? Are there other students in the content class that you should be comparing your content to?

Community and content audits

Let's envision a community audit as part of the content audit. Content strategy processes typically start with a content audit. A content audit for web content contains a listing of each page on a website and includes metadata about that page such as date, audience, purpose, and so on. For community content this step can be very messy. You may have to deal with multiple sites or with questions like, "do I count long comment threads in the audit?"

Also, when authors are not anonymous, you may need another axis upon which to analyze the content. For example, if a particular "rock star" blogger gets lots of hits, is it due to his or her real-world presence or the content itself? With sites like Klout, which shows how influential Twitter users are and who associates with whom, you can start to detect patterns in connections and influence. I believe this information should be used in a content audit to measure influence and track networking patterns.

A spreadsheet for a content audit looks like a page listing with columns for each attribute you want to record for the page. A record for a community audit might start with a matrix but may evolve into a social map that shows connections between community members. Imagine using LinkedIn data to find out who are former coworkers, who went to the same schools, who still networks and recommends others, and when were connections made. Access to this data underlies the real value of LinkedIn and other social networking sites.

When building your community audit, I'm not sure that details of business relationships help you build a strategy, but the overall picture should certainly inform your decisions. You can ask and answer questions like, are people tightly interlinked, are there small groups or closed "cliques," do people all come from the same corporate background (and therefore possibly have similar approaches to hierarchy and communication patterns)?

A community audit can help you make decisions about the number of professional writers you need to bring in, how often the content should be updated by a community member, and whether people will be willing to step in and edit without "permission," among other coaching and editing decisions. Will the community subgroups require a style guide before even writing a word? Are they the type who ask and answer questions more than they'll write definitive or prescriptive texts? Are they accustomed to intellectual property rules that prevent rather than encourage sharing?

The Social Technographics ladder[44] can also assist with a community audit by mapping demographic data such as gender and age to the likelihood that the community member will participate in a particular social web activity.

If you don't have a huge budget for design and a CMS to match the corporate one, do you go it alone to make a site that serves your particular customers?

Let's face it, the real blogs of our time are written by real writers who stick to it and know the difficulty of sitting in front of a blank editing screen and putting words in the box. Writing is a discipline, and blogs that are successful have a lot of energy and discipline behind them. In her book, *Clout: The Art and Science of Influential Web Content*[15], author Colleen Jones cites the results of a Holiday Inn study about hotel-manager generated content performance versus professionally written and produced content for each local hotel site. Apparently the professionally-created content so outperformed the hotel managers' content that they didn't even write up the exact results.

When your goals with content strategy also intersect with community strategy, you will find content being treated like a business asset and community influence and interaction directly affecting your business's bottom line (see the sidebar titled "Learning from IBM: Interview" (p. 190) for a description of Lisa Dyer's findings from her seven-year journey with community-based documentation techniques).

Case Study: Community Content Strategy at Autodesk

Often tools and techniques that incorporate XML and Component Content Management (CCM) are in the content strategists toolkit. If your toolset uses XML, can you still incorporate social web techniques? The answer is a resounding yes.

Appendix C describes how Autodesk repurposes its DITA XML content to build learning communities using the MindTouch social platform. Here is a related interview, by Scott Abel, of Victor Solano, Senior Manager for Learning Experience and Frameworks at Autodesk.

Interview with Victor Solano of Autodesk

In this exclusive interview[2] with Victor Solano, Senior Manager for Learning Experience and Frameworks at Autodesk, we explore how the software giant has leveraged the power of the crowd, software standards, and content management to meet the fast-changing needs of its customers. There are many lessons to be learned here, including the fact that the ways technical communication professionals have created user assistance content in the past may not be the way we do so in the future. An opportunity for change? Certainly. A tactic that provides huge efficiency gains? Indeed. Food for thought for your support efforts? You betcha!

TCW (The Content Wrangler): Victor, thanks for making time to speak with me today. After our recent webinar (recording available/registration required[3]), I've received lots of queries about the topics we discussed. To familiarize our readers who haven't heard the webinar, start off by telling us a little about Autodesk WikiHelp?

VS (Victor Solano): The Autodesk WikiHelp[4] provides the global audience of Autodesk product users with a new type of socially-enabled help system that combines the best content authored by Autodesk staff with articles, videos,

---

[2] Copyright © The Content Wrangler. Reprinted with permission. Full interview with many more links at: http://thecontentwrangler.com/2011/10/25/in-search-of-operational-efficiency-a-discussion-with-victor-solano-autodesk/.

[3] http://www.acrolinx.com/watch_webinar/items/operational-inefficiencies-spotting-process-problems-in-the-content-production-lifecycle-467.html

[4] http://wikihelp.autodesk.com

tips, comments, tags, and other content contributed by the vibrant and talented Autodesk user community. Currently, 26 Autodesk products and Suites publish their help content to the site in 14 languages. Many of the products use it as the default help system and have a context-sensitive help integration directly from inside the applications to the relevant content in the WikiHelp. All user content goes through a content moderation process to insure quality.

Autodesk WikiHelp content is continuously changing based on feedback, direct user edits, and a rapidly developing two-way conversation with our users. The experience combines help topics with learning videos, tutorials, and even a search tool that searches other Autodesk sites with learning content, such as our Discussion Forums, the Support Knowledge Base, Community Sites, and even YouTube. The system has direct connectors to the Autodesk Help Content Management System so we can publish content as an output directly from DITA XML to the learning portal and we can also batch-load videos.

The video on the Home page of the WikiHelp[5] explains the whole concept in 2 minutes. Make sure to expand it to full screen when viewing.

TCW: Why did Autodesk create WikiHelp? Why was it needed, exactly?

VS: Autodesk is an innovative company and this project is one of many experiments to try new approaches to improve customer learning. Since Autodesk products continuously evolve to provide a competitive edge to our users, it is important that our learning solutions adapt, keep up with the times, and recognize how our users prefer to learn. Given our users' preferences for visual learning, combining text and rich media, and a combination of hands-on learning

---

[5] http://wikihelp.autodesk.com/Inventor/enu/community/Videos/WikiHelp_Intro-duction_Video

options with non-linear just-in-time learning in a socially-enabled world, this project is an important step to better understanding how we can better serve those needs.

Autodesk WikiHelp seeks to enable an already existing community with new interactive tools that accelerate their learning process, rather than create a new community. We realized that many of our users were already consuming and contributing learning content in a vast and varied learning ecosystem with many competing options. Over 10 million Autodesk users are actively seeking and exchanging knowledge on Discussion Forums, Community Websites, AUGI (Autodesk User Groups International), blogs, Facebook, Twitter, industry websites, YouTube, and a rapidly evolving multitude of CAD and social media sites. They are connecting and providing each other with information, resources, and support. Some users expressed sometimes being overwhelmed with so many options.

Using the Autodesk WikiHelp, users can have access to the best help content provided by Autodesk, user contributed content, and leads on other resources in the community, with the ability to rate and comment, all in one place. It is a fresh new experience.

TCW: Why didn't you just improve your existing Autodesk help products and documentation? Isn't that the real problem?

VS: Autodesk has been winning Society for Technical Communication awards for help and documentation for many years. Improving our content remains our top priority. Still, user expectations change rapidly, and with the advent of YouTube, Google, fast access to knowledge, the social media phenomenon, and participation in a dynamic and connected world where users simply expect that their voices will be heard, the bar has been raised for everyone.

Autodesk WikiHelp has helped to us to offer new types of content, by combining text with rich media, integrating user-generated content, and has helped Autodesk products to provide a more continuous and responsive learning experience. Autodesk WikiHelp provides integrated HD-quality learning videos as well as topics, articles, tips and tutorials. Some of the top content may actually be contributed by our expert users, who never stop impressing us. Users can then rate and comment on the content.

In addition, there is a large need for specific niche knowledge that is best served by enabling peer-2-peer interactions and collaborating directly with our users, rather than exercising traditional methods. This helps us recognize and serve the Long Tail of Knowledge, a vast and shallow pool of specific user needs that tends to be underserved with traditional approaches and one size fits all solutions. Users tend to congregate and find each other as they have more specific needs to directly apply our applications. By providing integrated social learning tools with our Help, our users can help each other and we can collaborate with them to better understand and serve these needs.

TCW: Why do you think a socially-enabled help community like Autodesk WikiHelp is worth the effort? What are the benefits? How do you know your members find it as valuable as you say it is?

VS: We listen to our users and the response of the community has been phenomenal. Usage and engagement has exceeded the expectations of the products that are participating in the project. We continuously receive feedback on the Help pages through comments, emails through the Feedback link, user research, and we also keep a sharp eye on the site analytics.

The main benefit is improving the learning experience for our users. We do this by providing a more dynamic and continuous learning experience that leverages rich media and provides new tools[6] that allow Autodesk and the rest of the community to collaborate, making the whole experience more responsive and effective. Some other key benefits are access to detailed analytics and intelligence on content usage (see popular pages[7]), information on user focus and priorities, specific information on user learning preferences, as well as direct and specific user feedback on our content. This allows Autodesk to make better decisions to serve the user learning needs.

We have also seen an evolution of the roles of our writers and subject matter experts as they become increasingly engaged in direct interactions with our users.

TCW: Aren't there problems with allowing users to create content on the Autodesk WikiHelp site? After all, they don't know your style guide rules and most likely don't have experience as professional technical writers. Doesn't this approach just create a big mess of content of questionable value?

VS: Quality is supremely important, which is why we have a moderated system with some checks and balances. It can be challenging for an organization to allow highly edited professional content to exist side by side with user generated content and this can give way to fears that the users may "ruin the existing content" or "contribute content of questionable value." In practice, we have found the opposite to be the case. We are continuously humbled by the excellent quality of the content contributions of our talented user base as well as their ability to provide useful feedback for us to improve the content we produce. The big mess has failed to materialize and we have excellent content examples to show in its place.

---

[6] http://wikihelp.autodesk.com/training/enu/Contributing

[7] http://wikihelp.autodesk.com/Special:Popularpages

Further, all contributed content is moderated internally prior to publishing, weeding out any spam or content that does not meet our Wiki Guidelines and Terms of Use. After being published, it may be further edited by Autodesk staff or others in the community if deemed necessary. We aim to keep quality high by collaborating with our users to meet those goals.

TCW: What do you do to encourage participation in the Autodesk WikiHelp community? Do you provide incentives? Digital candy? Badges?

VS: On the home page of WikiHelp,[8] you can see a leader board with the top contributors. The system tracks contributions (edits, new pages, videos, and so on) and we can run reports to understand who our top contributors are for different products. Some products run contests and users register to win prizes.

In addition, our community leaders, bloggers, and social media folks invite and promote participation to insure visibility of the project and encourage engagement.

TCW: What types of roles did you create to help ensure the community doesn't become a vast wasteland of user-generated content? And if you created new roles, did you create internal roles (like community manager) as well as external roles (like super user)?

VS: Rather than create a multitude of new roles, what we actually did was to evolve current roles and processes and connect the users assistance organization to others across Autodesk. Some of more technical folks in the user assistance organization that are in charge of build and integration, quality checks, and production of our "help systems" to HTML Help or other formats, now also create WikiHelp outputs from our content management system (CMS). Some of

---

[8] http://wikihelp.autodesk.com

them serve as Wiki Administrators and help manage user permissions, create new product spaces, and help insure the system is running smoothly.

Now, in addition to authoring content, our authors and subject matter experts help moderate user content and respond to user inquiries through comments and the feedback link. Also, content authors may move some content from the WikiHelp back to the CMS system and edit it so that it can be integrated into the official stream, quality-checked, and localized according to our style and quality standards. To some degree, some of those roles have evolved to curate as well as author content. Our user assistance teams now collaborate more closely with the community teams in marketing, the web, and support. They have undergone training to understand new social media standards for interacting with the community directly and have become more connected with our social media teams. Some leads have taken training in analytics. They run reports to measure usage, engagement, and contribution and bring back this data to their teams to inform future decisions, measure progress, and help set new goals.

In regards to the "super user" question, we do also have some pilot programs with community moderators in our international communities that would fit this notion. We have worked with our Autodesk Community managers to reach out to some highly engaged users and community leaders. We have "deputized" some of these community leaders and top contributors with special super-permissions to moderate community content before it gets published and collaborate with our internal staff. They are also leading programs to help build user engagement in their communities and contribute top content.

TCW: What is the biggest value Autodesk has realized by moving to a Help 2.0, socially-enabled support community?

VS: Our focus is on creating value for our end users. By providing a combination of Autodesk authored content with access to knowledge from our expert community and enabling a multi-directional conversation to evolve, the learning experience has become more responsive and effective for our users. There is great value in responsiveness, providing continuous learning, access to new types of content, enabling rich media, and flexibility. We are also learning what works best and we have better tools to inform us of how to invest our energies in the future.

TCW: What changes – improvements, new features, innovations – do you expect to introduce into Autodesk WikiHelp in the future? What do you see in your crystal ball?

VS: We are continuously listening to user feedback on how to improve our learning options. We are gathering data from our users to inform our priorities for future development. We will keep you posted.

TCW: Some technical communicators and e-learning professionals have said that this new approach at involving the customer (especially in socially-enabled ways that allow user-generated or co-created content) is just a fad that will soon disappear after companies like yours realize that it was a bad idea with little value. What do you have to say to the naysayers?

VS: I would say "nothing ventured, nothing gained." It was time to take the conversation from an academic "what if" discussion to testing the waters and measuring tangible results. The early results are impressive and we will be continuously measuring value and progress. It may be that the data and results from this project help lead us to identify even better ways to serve our users' needs.

At this time, the Autodesk WikiHelp project is part of an effort to provide a competitive edge for our users and the associated feedback and hard data is very encouraging. We intend to always find the best ways to serve our user needs and that means taking action and sometimes defying conventional models to pioneer new ones.

TCW: Wow, it looks like we've run out of time. Thanks for sharing your time and valuable lessons learned from the Autodesk WikiHelp project. I can't wait to see how you are doing this time, next year.

VS: Thanks, Scott. It was my pleasure.

[Note: Autodesk WikiHelp is powered by MindTouch software. Learn more at http://www.mindtouch.com.]

More about Autodesk

Autodesk[9] is a leader in 3D design, engineering and entertainment software. Organizations in manufacturing, architecture, building, construction, media and entertainment industries – including the last 16 Academy Award winners for Best Visual Effects – use Autodesk software to design, visualize, and simulate their ideas before they're ever built or created.

---

[9] http://autodesk.com

8

Analyzing and Measuring Web Techniques

*A hundred objective measurements didn't sum the worth
of a garden; only the delight of its users did that.
Only the use made it mean something.*

—Lois McMaster Bujold

Let's discuss the processes, potential wins, and possible shortcomings
of web analytics and using the web for technical communication.
You always have to justify the cost when planning resources for your
department or for your job. Here are some ideas for managing and

championing these techniques using alignment with business objectives and tracking the correct measures.

Managing community methods

When introducing any new technology, you should ask yourself
whether it is necessary. What criteria make this conversation-based
implementation important at this point in the product's lifecycle?
How can you make it a priority?

You need to be able to write a business case for the cost savings or
time savings that could occur based on the conversation-based implementation that you are considering. Find out what business areas
need the most work right now. Understand your department's place
in the organization – are you more helpful to marketing and pre-
sales efforts or to technical support? Does your audience consist
mostly of internal consultants, and if so, what are their biggest
hurdles? How could conversation-based or community-based content help them meet their goals?

Also consider that your manager has access to more information
than you do and may understand the current customer climate and
community better than you do. Without the necessary tools and
information, you cannot easily build relationships, and relationship
building should start with your team's and manager's buy-in.

Convincing your manager

If your manager does not see any value in implementing social web
techniques for your content, continue to try to show the value of it
with examples from other companies or other areas of your own
company until you can make a business case that speaks to your
manager. If you strongly believe that a conversational approach is
the right direction for your documentation, try some grassroots efforts yourself until you can prove its worth and value, either with a

few great anecdotes or hard numbers that prove the return on the time invested. Also realize that *ROI* can stand for Risk Of Inaction. If *not* starting a conversation is more risky than starting one, that perception may sway your manager to try out a pilot project and experiment with social bookmarking, forum moderation, screencasting, or other areas that make sense in your business environment.

Part of implementing your chosen solution for integrating conversation or community into your user assistance is also recording lessons learned (what worked well, what did not succeed) and analyzing what you might adjust. Be sure to present metrics for measuring the success of your implementation.

Allocating time

All questions about using social media technologies distill to the single question, "How much time does all this take?" Beth Kanter[1] has some ideas. She says that each rung of the ladder involves more and more time. Expect to allocate 5 hours a week to listen, 10 hours a week to participate, and 20 hours a week to provide a platform for sharing content. For community building, expect the tasks to consume at least 25 hours of a team's work week. You likely don't need to add the listening hours to the participating hours once you are building a community, though if each of those tasks remains important and gives good return on the time spent, then the time adds up to sixty hours a week.

These time estimates may add up so that you realize you need more than one person assigned to this type of effort. In short, these tasks can easily become a full time job. The size and growth rate of the community will also dictate the number of hours needed to address their needs. Compare the time needed to welcome two or three people to an online community per day to welcoming over one hundred new members each day.

---

[1] http://www.bethkanter.org/

Measuring effectiveness

Metrics and social media mix like oil and water with interesting lava-lamp-like configurations as a result. In a corporate environment, someone has to pay the blogger's bills or the podcaster's bandwidth, and more importantly, decide that it is worth giving up other deliverables to spend time working on social media. Effectiveness, return on investment, and metrics come up continually.

When you are trying to measure effectiveness, be sure that you are using the right metric. For example, instead of measuring the Return on Investment of blogging, gauge Reach and Influence (R and I). Influence rather than Investment offers measurable analytics to use when investigating a blog's contribution to the bottom line. If you are seeking an independent blogger to sponsor or give ads to, you may use these measures. Use the same principles to analyze your corporate blog and measure the effectiveness of corporate bloggers.

An article by Ann Holland on the Marketing Sherpa site, "How to Calculate a Blog's Reach & Influence – More Complex Than You Think"[61], provides the following factors to use when evaluating and measuring the effectiveness of a blogger's influence:

- **Traffic:** Visitors and page hits, but don't trust traffic alone.

- **RSS feed subscriptions:** Subscribers measure return visits and may indicate loyalty to the blog.

- **Inbound links:** How often do other bloggers or other websites link to this blog?

- **Search position, part one:** Keywords for company brand.

- **Search position, part two:** Keywords for the blogger's brand.

- **Voice:** Does the person blogging represent the promise of an online experience that you can "hear" and appreciate?

I would add the number and frequency of comments. And the variety of the commenters' voices can show the type of audience the blogger has gained. When you measure these factors, rephrase Return on Investment (ROI) to be an investment in influence. That is, can you prove that influence is worth investing in?

After you have decided to invest time and resources to increase influence, how do you find the time for this "extra" work? Is it really extra? You can shape your role by studying the desired outcome and the values you place on influence. You may find the role of enabler or curator yields better results than content contributor.

If your measurements prove the value, showing time or costs savings should be straightforward. Look for sales or connections resulting from a blog entry, wiki article, or customer support post. For example, compare money spent on a marketing and public relations campaign with a few hours to write a blog entry.

And how about the opposite question: if everyone else writes the content, what do writers do with the time saved? Edit like a magazine publisher? Become a content curator or community manager?

Rahel Bailie, content strategist and founder of Intentional Design, believes this is the new role for technical writers, saying in the article, "Managing Online Communities: The Next Big Idea for Communications Professionals?"[42]:

> To be able to leverage UGC,[2] an organization needs to have a community from which to draw information and opinions, and those communities need moderators. Community management, a relatively new career track, seems to be a logical next step for content developers.

---

[2] UGC = User-Generated Content.

Fitting into the community

You should try to match your contribution with your role in your company. For example, if you add value by decreasing customer support calls, then look to the support communities for ideas. If you assist the services or field personnel in getting their jobs done, find a way to tap into that community base. If your documentation is a sales-winning essential part of the product, get close to the sales and marketing department's conversations with customers.

Matching your strengths and experience to the conversations helps you avoid stepping on toes or stumbling into conversations where you do not have the tools, background, or correct messaging to handle the situation. Other departments may be able to help if a customer becomes angry or you witness an outburst between two community members.

Clay Shirky has an excellent essay, "Group as User: Flaming and the Design of Social Software"[80], which describes cooling-off techniques, rating, karma, and ways to harness the "performance" effect for good purposes.

Brand protection, public relations, and crisis communications are best left to trained professionals.

However, brand protection, public relations, and crisis communications are best left to trained professionals. You should find someone at your company who can help in those areas. You should also be ready to escalate the information to the right group when you observe a conversation that another organization should handle.

Encouraging grassroots efforts

Grassroots efforts are defined by an unofficial group starting a pro-ject without authority or hierarchy but with organic growth and natural networking tendencies. When a few friends or coworkers who share a common vision or goal start on an effort that grows beyond their expectations, that would be a grassroots effort.

Another term that is used rather than grassroots is "groundswell," from the book by the same name. In this context, a groundswell is:

> ... a social trend in which people use technolo-gies to get the things they need from each other rather than from traditional institutions like corporations.
>
> —*Groundswell*[19]

It is also described more simply as a change in public opinion that is broad and far-reaching. Writers can build a groundswell because they know the product well, can help people make connections with each other using social technologies, and can write well enough to convince others of the basic value of the product. Writers may also already know community members who might be willing to help with a few tasks or at least be willing to review write-ups. With a publishing platform like a blog or a wiki, writers enable conversations to take place that may not have had a place to occur.

Writers at a mobile phone designer and manufacturer who wanted to experiment with a wiki decided to go through the correct channels for a legal disclaimer by simply requesting revisions to the legal verbiage in their existing online help disclaimer. They were pleasantly surprised when their legal department was helpful and encouraged their experiment. Then the writers created a MediaWiki installation, pre-seeded it with content from their user guide, and waited to see if users would find it and edit it.

Interestingly, a few years after this experimental wiki started, it turns out that another wiki, which was built by end-users of the cell phone they intended to document and build a community around, became "the" source for information and user-to-user communication about the phone and tips and tricks.

In that case, while the company wiki was helpful, the external community wiki was more visited and edited. I think their experiment showed their relative naïveté about the power of community building rather than web page building.

Another term related to grassroots is "astroturf," which is a group that wants to be known as grassroots but is conceived, created, and sometimes funded by a corporation or lobby. Full disclosure of the backing behind your efforts is the best course of action to avoid distrust or suspicion later when someone reveals the origin of the group and the underlying motivations.

Recruiting others

Once you have a few success stories to tell, share them with anyone who shows an interest. Storytelling is at the heart of a grassroots movement. Blog entries and interviews with wiki maintainers are examples of storytelling that you can do within your organization.

At the "Social Media Metrics, Where are they?" panel discussion at SXSW Interactive in March 2008, Ynema Mangum presented a chart showing the CXO "X-factor" for influential metrics for each area of a large business, where the X might be replaced by Operations (COO), Technology (CTO), Finance (CFO), and so forth. The X-factor is their business area of expertise and indicates their role in the company.

Table 8.1 – CXO "X-factor"

X-factor	Influential Metrics
Executive – CEO	Aggregate of all metrics you can collect: reported in an interpretive way, with the goal of getting executives involved using language that matters to a CEO.
Marketing	<ul><li>Impressions vs. Engagements: converting someone from a prospect to a lead to a customer.</li><li>Number of sales leads</li><li>Time spent</li><li>Sales lift</li><li>Payment for "buzz" generated by a press release can be directly measured. Agencies may charge more than $800,000 to ensure that their "conversation" gets on the front page of a site like digg.com or slashdot.com.</li></ul>
Communications	Measurement of the number or conversations: measured with a negative or positive tone or overall effect. Sentiment marketing: measured by the number of positive, negative, and neutral comments.
Finance – CFO	Value they are getting out of each channel: email (clicks, opens, and forwards), mentions, and so on. Cost per conversation or revenue per conversation.

X-factor	Influential Metrics
Technology – CTO	Implementation, security, and integration.
Operations – COO	Efficiency
Human Resources, Administration	Retention rates and attention from potential recruits.
Sales and Business Development	Revenue/growth

Once you have determined what you'd like to achieve, consider the direction of the conversation and how many participants it will have. Sarah O'Keefe at Scriptorium Publishing has an excellent white paper titled "Friend or Foe? Web 2.0 in Technical Communication"[75] that describes the one-to-one, one-to-many, many-to-many, and many-to-one directional communication occurring today.

If you want to encourage generated content, start slowly. You might seek business objectives at different levels, depending on your goals and the business areas you aim to influence.

IBM community strategy

Learning from IBM: Interview

Lisa Dyer works at IBM as a community strategist. She has found many connections between open, community-oriented documentation, fulfilling the needs of the business, and serving customers.

She worked for Lombardi Software, a company that was acquired by IBM in 2009, where she worked on converting DITA content to Confluence wikis and built a community around professional service providers.

Her division of "warranted" and "non-warranted" content was described in the section titled "Differentiate "warranted" content" (p. 101).

This interview takes a look at her perspective in mid-2012.

1.   How are you aligned in the company? Support, marketing, training? What is your job title?

My job title is "Program Manager, Business of Community" which is a role embedded in Product Strategy.

My journey from Engineering to Strategy was a long but intentional one. It makes sense, given that the business of community directly supports the top goal of any line of business: growing the business.

A community strategist can help drive grassroots synergies, scale an organization's ability to self-enable clients, and effect cultural change in how you support and leverage the community.

In sales and post-sales interviews, if you ask how important is information to you, you have either "very important" or "important" as the response. You'll see almost 90% in these distinct areas:

■ **Purchase decisions:** Determines whether they buy from you or a competitor, they are influenced by quality and availability of information.

■ **Perception of the company:** This could be brand perception, could be quality perception, how do you view the brand based on the information that is out there? Is it a mature brand? Increasingly, being mature incorporates having a vibrant community.

■ **Continued product satisfaction:** Does your info self-enable customers or do they have to go through a ticket route?

■ **Product quality:** Do they recognize their users need information? Are they investing?

This is not new, it's treating information as a strategic asset for any company. Also, it's not enough to define goals, you have to have key performance indicators to measure whether your goals are being met.

Find key performance indicators that you can roll up all the way to the top of the value chain or your organization.

2.   **What types of content do you deliver?**

How-to recipes and cookbooks for solution architects and developers, methodology and best practice, toolkits and sample apps, forum Q&A, API reference and other tech docs.

At first, the community wiki asset was available only to customers. Now it's free and open to a worldwide community. It was a change in mindset – that we thought the competition would take the content and take it and reuse it, even resell it with their services, but the shift came when we realized it's not really the information trails, they get the docs regardless of your tracking, you don't really have control over that.

It's the community creating the content and having those valuable conversations that gives you the competitive edge. Let me emphasize that the upsides really outweigh the downsides – the benefits of sharing outweigh the risks nearly every time. And you can't control all the messages anyway.

3.   How is your content licensed and how was the license selected?

Under the ibm.com Terms of Use,[3] which explains how the content is used.

4.   How many collaborators do you work with regularly?

Dozens of people across each type of content. We recently published a 170-page RedBook, with a group of about recruited experts, they don't have to be IBMers. Usually such an event and output occurs in about 5-7 weeks, we fly them into the central location so it's quite efficient, time-wise.

5.   Do you hold sprints and if so, how many people are active in sprints?

We use sprints to refresh content, sprints to reorganize content, sprints to develop new content. Participation varies from two people to dozens, averaging around 5-8.

I host community calls now internally, to transfer knowledge and share within the company-only. The sprints are a very effective knowledge transfer as are the calls. We do a sample exchange every other week, developing sample assets to showcase new features.

We've developed almost 20 new samples through this information exchange. It's a focused effort with the community.

We also have a wish list to find out what people want to know how to do – even "how to capture errors in a custom log file" is one item on the wish list.

---

[3] http://www.ibm.com/legal/us/en/

6.   How many page views do your pages get?

The page view numbers give you a good sense of scale (in our case, 120K + per month). But I think the more interesting Key Performance Indicator is this: the average member logs in every weekday and stays for 9 minutes. This observation is an important part of proving the value of the site in a user's daily routine.

By focusing on what keeps people coming back and spending more time on the site, we scaled to 55% more page views and 28% new memberships in just one year. The community self-enablement strategy is working.

7.   What is your tool chain and how is it created? Internally, with outsourcing, through the community?

It's a commercial wiki that has a lot of built-in power and an extensible plugin architecture. A wiki should give you the agility to constantly update and hone the user experience with the website. We ensure it gives good findability and visibility. It's important to find a way to break down the information silos caused by multiple systems of record (sources), and give the customer a single entry point as part of this good user experience.

8.   Anything else you want to share about your particular situation?

First go-live in 2006 (entitled users only), followed by open access in 2011 (as of this writing, "open access" means there's a required, but free, sign-up process so users can view content.) We have "warranted" and "non-warranted" content. Comments are available only on "warranted" content, the Edit button is available on the "non-warranted" content.

> Both types (with a 60-40 split in favor of non-war-ranted content) get pointed to from the forums and has more pageviews, answering particular questions. You're showing "this info already exists, no need to duplicate it" plus the fact that people are finding the info and pointing to it. It shows the degree to which the community makes sure the link gets the attention it deserves.
>
> I'd also like to point out the cultural transformation you can drive with your community... you will meet resistance with the nature of non-warranted content, but your message to resistance should be this: Don't be afraid to embrace community-developed content. Used responsibly, it can dramatically accelerate your projects and scale your knowledge, and the goodwill and battlefield experience of the community is uniquely valuable. Everyone is freed up to deliver more value, innovate, and shape their careers.

Measuring documentation as conversation

This question, "But does documentation as conversation work?" stems from the measurability equation. Those who are interested in website metrics and analytics often want some fancy tool to take care of the measurements for them. But in reality, it is extremely difficult to know what measuring stick you can use on a blog entry, for example, or the value of a well-stocked wiki. Measuring community involvement is often discussed in fuzzy vague terms and not hard numbers. So, if you apply conversation or community methods to documentation, how do you prove it is working? And how do you prove the level to which it's working and whether it outpaces conventional user assistance methods?

Website analytical tools like Google Analytics help you count numbers of visitors, and the tools are getting more sophisticated about measuring how long each visitor stays. Also, they are better able to track the path that users take through a help system, which can help you measure their effectiveness. Plus, Google's keyword tracking is extremely helpful for indicating which keywords are most searched for and may help you analyze which keywords are most efficient for letting a user figure out the answer and move on.

Time spent on a page may or may not be a good measurement for effectiveness of a help system, but findability of your help system is measurable. You can also ask your audience if their participation in conversation or community has prevented them from leaving a company or abandoning a product from a particular company. Often we don't measure in the sense of preventing a negative result.

Basic web analytics measures include the following:

- Page views
- The total number of page views in a specified period.
- Unique page views
- The number of visits during which the page was viewed.
- Exit or Bounce rate
- The percentage of site exits that occurred from the page.
- Time on page
- The length of time that a visitor stays on a page during a visit.

Community equity is an emerging research area that measures the value that an individual brings as a member of an online community.

At Sun Microsystems, Peter Reiser developed a method and al-gorithm for calculating a member's community equity based on in-formation offerings and personal interactions with others. According to the website, a community equity framework helps to build a social value system for social networks and communities. The framework is written in Java and Enterprise JavaBeans (EJB).

His team also created widgets that help people see other members equity ratings and central areas of expertise. By reading, rating, contributing, tagging, and editing, a person increases their inform-ation equity. Personal equity is calculated based on connections, reputation, and collaborative efforts. The values "age" and decrease to zero over time so there is a "freshness" to the values. Peter Reiser's project is now available as an open source project.[4] These individual visual measures of a community member's contributions may help you measure the overall value of a community.

In the 2008 SXSW Interactive panel presentation titled "Social Media Metrics: Where are they?" the producer for talk.bmc.com community content, Ynema Mangum, noted a measurement case study of a single blog entry written by William Hurley (known as Whurley) with reasons why Microsoft loves open source. Whurley's blog entry was written to draw attention and succeeded. It was featured on the front page of active online communities such as Slashdot and Digg.[6]

It could cost $800,000 dollars to gain similar media coverage. That is one method to measure conversations, though Whurley's use of conflict journalism and sensationalist shots at competitors does not fit nicely into our usual methods of delivering quality information in a timely manner.

So, how could you measure whether a technical publication-based conversation is working? My first suggestion is to go to the usual

---

[4] http://kenai.com/projects/community-equity
[6] In an ironic twist, despite its notoriety at the time, the original article, "Seven Reasons Microsoft Loves Open Source," is no longer available on the web.

metrics that technical publication uses to provide value, for example, lowered support costs. Or, perhaps it is not Return on Investment you need to measure but Risk of Inaction.

If your competitors all use wikis or help systems with comments and other conversation-based, community-based user assistance tools, your team should evaluate their use for your deliverables as well. Just as competing products need to keep up with each other's features, I think that we need to keep an eye on competitor's conversations and determine if there's a perceived value your company should match.

Let's expand upon the processes, potential wins, and possible short comings of web analytics for technical communication based on my experiences with several sites tracking data over years. When I spoke with a few Google technical writers at the STC Summit in 2010, one of them confirmed that their performance reviews include a web analytics component.

This concept got me thinking about help sites I have worked on and how well they would stand that test. Or rather, how well my writing and information architecture would stand up to an investigation with web analytics data.

What are the business goals?

I believe tech pubs groups may serve different masters or several masters. Pre-sales or marketing goals are different from support goals, and training or education goals are different still. So you should pay attention to different measures depending on your goals for the site.

Let's take the example of customer support and consider the goal of avoiding costs caused by support issues being filed. The goal here is preventing support calls. In *Web Analytics Demystified*[24],[7] Eric Peterson points out a distinct difference in goals for a customer support site, saying:

> while the other business models are driven financially by the idea that more page views are usually better, the customer support model tends to be the opposite, the more quickly the visitor can find the information they are looking for the better.

Peterson's book helps us understand how web analytics tie into business goals. The book has an entire section devoted to customer support sites.

For training sites, the goal would be for the person to spend time with the content, digest it, and meet training objectives. Time spent by a web visitor should be higher for training sites. Another crucial difference between a support site visitor and a training site visitor is that for the support site visitor, you want to observe the behavior of new visitors, but for the training site visitor, you want to ensure retention and repeat visits.

To put it simply, customer support deals with acquisition of visitors, training deals with retention of visitors. Another business goal is conversion, converting site visitors to paying customers. A technical manual can assist with three main goals: acquisition, retention, and conversion. You can how the conversion by tracking funnels, the steps through web content that a user takes to get to a goal.

---

[7] I downloaded and devoured two web analytics books from John Lovett and Eric T. Peterson on their site at http://www.webanalyticsdemystified.com. The titles, *Web Analytics Demystified* and *The Big Book of Key Performance Indicators*, are available for a free download after you give them your email address.

- **Acquisition.** Gathering new readers, getting readers to bookmark and share your site.

- **Retention.** Ensuring readers return to the site.

- **Conversion.** Get a reader to do a particular action (not just buying a product), such as downloads, taking a quiz, gain understanding, complete a tutorial.

So how about support cost deflection? What would you measure? If the site has a question and answer section, compare the page views of the FAQ or Q&A pages to the other pages in the site, are there more views and longer time spent on pages in the FAQ area? That might be a good sign to indicate the help site is deflecting support calls.

Search analytics are another very important area to study to find out the effectiveness of your web content in achieving the main business goals. Also you should keep an eye on search terms that yield no results. Check for content that is missing and terms that your users seek that is not a term in your assistance.

These analytics can be very powerful for showing the "helpfulness" of your help site. For example, the Mozilla support site added a "Was this page helpful?" rating on each page of their open source help system. They had remarkable results. The page titled "How to set the home page" went up +13.1%, serving 1,170 more people per day. The page that described their Profiles went up +33.6% (an increase of 1,122 people/day). With this data, they could prove they helped 800,000 more users in a year based on higher ratings after a re-write. The effort involved in the re-write was paid back in proven increases helping support their users.

What to configure

Set up a custom segment to look for pages that have troubleshooting or particular error messages in the title or page content. Next, look at the bounce rate for that segment compared to the rest of the site. You want the bounce rate for the troubleshooting topics to be lower than the overall bounce rate. You want the trend for bounce rate for troubleshooting pages to stay the same or go lower over time. In other words, if visitors do not spend any time reading the troubleshooting information you have provided, what can you do to improve the content to prevent a visitor from leaving (bouncing)? This screenshot shows an example of a lower bounce rate for the troubleshooting segment of pages compared to all pages, which you want to maintain if your goal is to help users troubleshoot independently.

Pay attention to the pages that have the highest rate of exit, the page that most people leave the site after viewing. But also look at the percentage of exits. I have seen great pages on tech comm sites that have 9% exit rates but the highest number of page views. One explanation is that non-users found them through organic searches and the entire site was not what they expected, not just that page.

What is the responsiveness of people to updates or comments on the site? If it offers comments, how much time elapses between a comment and a corresponding response? Do the questions get answered by a company representative or by another user? Not all web analytics packages will offer this measurement so you may have to do a sampling yourself. You can also try to get a sense of cadence, the rate of comments per day or per week or per month.

Analyzing searches will go a long way towards understanding whether user's needs are being met by the help site. Look for searches that have zero yield, that is, the user did not click through on any hit or no results appeared at all. You can also look at the search results to site exits ratio.

Watch first time visitors data like a hawk. New visitors may struggle at first while they learn their way around the site. As the number of first time visitors goes up, support call volume may also increase if visitors cannot find what they need or if they find the call-in number quickly. A good method for tracking real-world data along with website data is to use a special phone number displayed only on the online help site, so that you know only those people who found the page with the special phone number can call it.

You can set up a visualization funnel from the home page to the support site to specific information to generate what Eric Peterson calls the Information Find conversion rate. For example, consider a flow of visits to the home page, to a product page, to a list of commonly asked questions, to a page containing a specific answer to a question. You can track travel through this series of pages, measure abandonment along the path and track a conversion as a certain amount of time spent on the final answer page.

Where analytics fall flat

Analytics might paint a picture or help you tell a story, but they tell you only the what, and they do not explain the why. You must do the analysis about why the data is affected in this way and make decisions based on the data. You may have two few visitors to make any decisions, or you may need to collect data for longer to ensure you have good before and after comparisons. Also realize that the technical communication industry has very few benchmarked successful sites to compare your site's data to.

Limited data will certainly make it harder to provide a convincing, statistically-significant analysis. One of the problems I foresee with applying web analytics on technical documentation sites is the small number of page views per day. I recently analyzed a help site that had about 40-50 page views per day on weekdays, pretty consistently. That was a software-as-a-service product available online. Another site I have watched for more than a year consistently gets 200-300 page views on weekdays and just under 100 page views on weekend days. I hear but do not have an official citation to point to that the 10k visitors per month is a common benchmark for starting to pay attention to web analytics. Do we get much value or accuracy from analysis if our sites do not reach that point? I think it is okay to monitor but to recognize your data may not have the clout you would like it to.

Benchmarks to compare your site to would be valuable, but there are no categories as specific as say, help site for consumer gadget, or help site for enterprise software, yet.

Site search analytics, while most valuable to us, may be harder to enable unless you use specific tools. Site search analysis shows you what users look for, whether they find anything, and the path they take after clicking on a result link. Search analytics focused only on your tech comm or online help site require you to use a Google Custom Search Engine or the MindTouch 2010 (and higher) platform which has site search analytics built into their reporting system. It appears that Adobe RoboHelp Server Analytics offers the ability to see what users search for but I do not know the depth of analysis beyond keywords.

Connecting to the greater web analytics group at your company may be a challenge. Google Analytics is the free offering, so I expect it would have the highest uptake in tech comm in the beginning. Also, tech pubs departments are not usually tied in to the web content management systems such as Omniture or Coremetrics (now owned by IBM) which are two other web metrics tools may not gather data on a tech comm site.

One takeaway from *Web Analytics Demystified* is the question, "Is the information actionable?" In other words, when deciding which metrics to watch, make sure you can do something about the resulting metrics, whether you make changes to content or dive more deeply into the metrics. In certain environments, actionable items could be problematic if the business is attempting to change, seek different markets, or when politics and power struggles internally cause controversy about actions to take.

9

Open Source Documentation

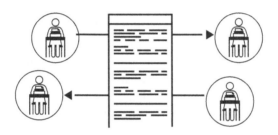

*If programming is like bicycling, documentation is more
like basketball. The best players don't always win.*

— Dana Blankenhorn

Open source, how does it work?

I'm often asked by writers, "What is free libre open source software?"
The first step to understanding is spelling out the acronym FLOSS,
which means you need to understand the meaning of free, libre, and
open. Fortunately there's a great simple explanation available in a

question and answer set on the FLOSS Manual's Questions and Answers page, describing both free and open:

> Open Source emphasizes availability of source code to software users.... Free Software emphasizes the freedom to modify and reuse software, which of course also requires that source code be readily available.
>
> —FLOSS Manual's Q&A page[1]

The keys to open source and free software include the ability to use, modify, and redistribute the software. These key attributes may or may not extend to the documentation for the software.

Free software advocates like Richard Stallman argue that free[2] software is not the same as open source software, and for developers, that is true. However, for most users, there is no significant difference between free software and open source software. If you are interested in the details, Eric S. Raymond's article, "Goodbye 'free software'; hello, 'open source'"[77], which began the movement to avoid the ambiguity inherent in the term "free" by replacing it with the term "open source," or the definition in Wikipedia for open source are good places to start.

Because people think of free as no-cost, expectations for documentation and support fall by the wayside with free software. Free software and open source communities are quite developer-centric rather than user-centric. There's an inherent conflict at play with every open source community – when priorities are made, do support and documentation rise above the basics of coding and debugging? Rarely, if ever, and only for the best funded projects.

---

[1] http://booki.flossmanuals.net/floss-manuals/questions-and-answers/
[2] I have heard that the word "free" is just a bug in the English language because there are too many definitions available for free.

On the Linux and Open Source blog on ZDNet, Dana Blankenhorn summarizes a post titled "Why open source documentation lags"[45] by saying,

> If programming is like bicycling, documentation is more like basketball. The best players don't always win.

He offers great explanations for the lags in documentation, and the reasons are not just tied to open source software – all software documentation could use more team sport and collaboration efforts to create decent documentation.

The biggest weakness for many open source software projects is the lack of good documentation, it really does delay maturity of the development processes and it also hinders wider update and adoption. Documentation is often a success factor for supporting software.

Reasons for the delay and lack of documentation are varied and really depend on the community culture, vibe, and heart. Sometimes there's a culture clash between developer types and designer types, where vitriol and "argument for entertainment" turn off potential non-developer contributors who simply don't want to participate.

Other times I've observed a lack of tooling, which can prevent design and documentation feedback loops from working well. Some see this "lack of tooling" as a weakness of documentation folks who don't want to work in source control, but I haven't seen this avoidance of source control by writers.

Dana observes that only final polished documentation is useful to users. Rough early drafts may be plain wrong and misleading, which is rather unfair to the documentation folks who try to publish early and often and iterate through drafts. I would counter that some documentation is better than no documentation, but I do see an inherent problem with that approach when accuracy is key. When documentation becomes untrusted, and people follow a procedure for an hour only to find out it won't work, the emotion about the docs turns into hatred against the documentation. No one wants that reaction.

Another interesting conundrum for writers of technical information for open source projects is that information and training are good ways to make money with open source. As Adam Hyde, founder of FLOSS Manuals notes in an interview from March 2009,[3] there is a known model to "Write the free software, then write the closed proprietary documentation for a buck and clock up some status." Reputation building is certainly a motivation to write documentation for a project, too.

So there is an inherent conflict when the software is openly licensed and freely available, but the documentation is closed source. There's also a known problem with people who identify themselves as the "authors" of the "official" documentation and in doing so effectively block others from collaborating with them. This phenomenon is not unique to the world of open source software. Closed software also has authorship difficulties where collaboration just doesn't happen because it isn't part of the values shared by others in the system. One author per "book" can also lead to this type of mentality.

Documentation everywhere suffers from a lack of "shiny polish" or "sex appeal." Web design and interactive design and experience design all help fight this perception, but there's still a ton of unsexy, boring documentation lying around.

On Network World, in a post titled "Creating a library of FLOSS Manuals"[88], Amy Vernon asks, "...why do so few applications have manuals to start with?" Her initial answer is tied into the use of manuals, asking her readers, "When's the last time you read a

---

[3] http://www.netsquared.org/blog/alexsteed/interview-adam-hyde-floss-manuals

user manual?" Fortunately, she found the offerings on FLOSS Manuals to be quite useful. And I think that's the key to software documentation. Whether it's open or closed, the usefulness of the doc, no matter what form it takes, will be its final measure.

Open documentation community strategies

Some open source documentation efforts are led by an all-volunteer work force, but I've found that there are several other models for open documentation. The models fit on a spectrum from an all-volunteer writing team to a mixture of paid, professional writers and community collaborators to all professional writers. For some projects, a single paid position whose focus is documentation pays off in much higher adoption rates and gives support channels high-quality documentation to aid in support efforts. Other projects work well with a strong all-volunteer core documentation team.

Translation efforts often fit on the same spectrum with some projects using an all-volunteer translation crew and other projects outsourcing all of the translation to professional teams.

How do collaborative efforts produce high-quality open source documentation? How does working on open source documentation differ from working in a corporate environment? Let's explore these two questions by studying a few documentation communities, namely FLOSS Manuals, Gnome, RedHat, and OpenStack.

FLOSS Manuals

FLOSS Manuals was founded in 2006 by Adam Hyde as a non-profit in the Netherlands through grant funding. FLOSS Manuals enables a community of open source writers to write high-quality documentation for free software through both in-person facilitation and collaboration tools.

The FLOSS Manuals site went live in May 2007, and the first community-based edit on the site was in July 2007. Many of the original manuals were written by Adam for his workshops about streaming media, since he is a digital artist.

Originally the collaborative software used to create manuals on the FLOSS Manuals site was based on TWiki, but now FLOSS Manuals runs on open software created by a small team called Booki. FLOSS Manuals has created specialized tools that enable book sprints as well as translations.

Their funding model relies mostly on volunteer writers, though some specific needs are commissioned using professional writers. Adam Hyde works on FLOSS Manuals full time and facilitates book sprints for much of his personal income. Other sources of income for the organization include collaboration with organizations in need of documentation, grants, and funding from organizations in need of tools. Some writers earn a commission for writing a manual, often funded by a group that needs that particular manual. Also the proceeds from book sales are invested back into the FLOSS Manuals organization to further the aims of the group.

Collaborative efforts at FLOSS Manuals are highly focused on the book sprint model, but an interesting secondary collaborative effort is the development work around Booki, the software that helps you make free books, in both printed and ebook formats. The work done at FLOSS Manuals is different from a corporate environment because of the non-profit nature, the volunteer workforce, and the focus on building tools in addition to building community. It's a unique group with highly focused leadership.

The book sprint technique used by FLOSS Manuals is described in the section titled "Book sprints" (p. 53). A recent experience for

FLOSS Manuals was a set of four book sprints at once at the Google Summer of Code Doc Summit. The four projects were KDE, OpenStreetMap, OpenMRS, and Sahana. Each project brought between three and five participants. Some of the participants didn't know they had applied to write a book in a week, while one project already had a long outline, reviewed by other community members.

The participants' backgrounds and statures varied widely – tall, wide Python developers with long hair in a ponytail, disaster managers with cropped hair and glasses, a geology professional with a travel bug, cycling data wranglers, web developers, C++ developers, young gamers, and a user advocate who also happened to be a grandma. And then there were the doc aficionados who find all of this quite fun, like me.

We started the week with an unconference led by Alan Gunner from Aspirations technology,[4] who has a way with words and a distinct gift for engagement. No one hid behind their laptops (though sneaking a peak at a small screen like a mobile phone did happen sometimes). About thirty of us were gathered, and I feel we built a community of practice that I can continue to interact with to talk docs. These connections as well as the resulting books were a valuable gift from Google, whose Open Source group funded the week with lodging and food provided.

OpenStack at Rackspace

OpenStack provides open source cloud software to organizations that want to build a scalable cloud for computing power or storage solutions. In the fall of 2010 I started working at Rackspace on OpenStack documentation as a community documentation coordinator. When I arrived there were just two projects, Compute and Object Storage.

---

[4] http://www.aspirationtech.org/blog/gunner

Most of the existing documentation was for developers, and interestingly, fell into two types:

- **Python developer documentation:** This was documentation for people writing code and it lived nearest to the Python code itself. These developers needed to understand how the code base was put together but did not necessarily need to use OpenStack.

- **REST API developer guides:** This was documentation for application and UI developers, mostly consisted of the REST API specifications, written by architects with a vision for the future of the API. The developers these are written for are consumers of the OpenStack APIs, developers who run large web sites with a need for streaming media, or developers who need large amounts of computing power in small bursts.

After completing a content audit of these two sets of content, I decided to segment the content based on source. The Python developer documents were written using reStructuredText (reST), an asciitext markup language that uses Sphinx to build HTML output.[5] The API guides were written using DocBook. I hoped that since RedHat and Fedora guides were based on DocBook that my future collaborators would like working with XML-based documents. I decided to build a documentation website for OpenStack cloud administrators, deployers, and consumers at docs.openstack.org.

Previously, the guides were only being output to PDF, and they had Rackspace branding. I needed OpenStack guides without Rackspace's voice and in webhelp, an HTML-based format.

As is often the case with open source projects, I found help in the community of DocBook users. Kasun Gajasinghe, a Google Summer of Code participant working with the DocBook project, and his mentor, David Cramer, helped me produce webhelp output.

---

[5] http://sphinx.pocoo.org/rest.html

With the help of Todd Morey, a gifted web designer, we made HTML output with Disqus comments on each web page. I announced the documentation comments at a cloud event after the second time-based release of OpenStack, and people were very excited for the chance to "talk back" to the docs.

However, the journey was a tough one. We started with a 66% bounce rate – two-thirds of visitors left the site almost immediately! I'm happy to report that we've flipped that one to the opposite – now two-thirds of visitors stay – and we've built a loyal return audience as well. In six months time, I was no longer the only one responding to comments on the site, and other people started answering each other's questions.

The docs.openstack.org website gets over 300,000 page views and over 27,000 unique visitors per month. The wiki at wiki.openstack.org, which houses project information more than end-user documentation, gets about 140,000 page views and 20,000 unique visitors per month. The developer sites such as nova.openstack.org, which are built automatically when the code builds, get about 28,000 page views and 5,000 unique visitors per month.

The OpenStack documentation team has about a dozen members from many participating companies. I get the sense that docs are written by people making a paycheck to work on OpenStack, but only a few so far are paid professional writers, and they do not work on OpenStack only.

I work on a team within the larger Rackspace organization called the "Cloud Builders". We're a community team that mostly aligns itself with adoption, support, and training. The title I chose for myself is "Content Stacker." I work mostly on a site at docs.openstack.org, delivering system administration manuals and API guides.

The OpenStack project has attracted over 140 companies with potential contributors, but the documentation collaborators are a small group. For the first OpenStack release, the documentation site had about six contributors. During the six month release period, there were 18 contributors, including a team of nine who wrote an entire book, the *OpenStack Compute Starter Guide*[22].

I'd say there are now fewer than 10 regular documentation contributors, other than developers who write code comments and explanations for features they add. But, the automation and review systems we have in place have enabled this small team to be very productive, creating 175 doc contributions in a six month release period.

I coordinate documentation efforts and stack content into meaningful bundles across multiple "core" projects that help organizations adopt OpenStack as consumers or deployers. I manage the documentation project just like a core code project, including doc bug reporting, task tracking, build tool debugging, translation, and all of the continuous integration tasks required to keep up with a fast-paced software project.

I also support the wiki, support developer sites that publish doc strings embedded with the code, customize the search engine, seek new content and new contributors, run a monthly doc meeting, participate in all Design Summits, and make sure the vision for the docs aligns with the vision of the project.

The vision is to increase adoption for OpenStack to help Rackspace meet its strategic goals along with HP, IBM, RedHat, and fourteen other member organizations investing in open strategies.

Our content is intentionally licensed openly. For developer documentation that lives with the code, we use the Apache License, version 2.[11] For manuals, we use the same license, but we let contributors offer content with a Creative Commons license.

---

[11] https://www.apache.org/licenses/LICENSE-2.0.html

For example, the contributor team – an open source training and consulting company – that created the *OpenStack Compute Starter Guide*[22] licensed it with the Creative Commons Attribution NonCommercial Share Alike 3.0 license.[12] I have also approached bloggers who use Creative Commons licenses and requested permission to insert their blog entries into the OpenStack documentation. This approach encourages contributions while honoring the original intent of the authors for reuse of their content.

I have held one one-day doc sprint, during a Design Summit – an event we hold every six months to plan the next release of OpenStack. I have found that doc sprints must occur with specific releases. Originally I thought I'd just run sprints at the Design Summit twice a year, but Design Summits occur after a release. Now I see that sprints should probably occur just prior to releases.

For the last release, I held a doc blitz instead of a doc sprint. It was a one-hour documentation review held world-wide with specific channels set up for feedback, allowing doc comments on a specific "docblitz" site, IRC, and Twitter. My email inbox had a comment a minute incoming at one point during the blitz!

Most recently we held a one-day doc sprint that was a doc day. I traveled to San Francisco from Austin, since we had about 40 people signed up from the local Meetup group. About 20 people dropped in during the day and we had great discussions.

I'm proud of our tool chain. I get to help shape the build tools and output tools. We also have a great review and publishing process that we continually improve. To work on the DocBook source files, we have a working project in Github in the github.com/openstack/openstack-manuals project. All API documentation has been moved to Github and is housed in image-api, identity-api, compute-api, and object-api projects there. For Github hosted docs, you get

---

[12] http://creativecommons.org/licenses/by-nc-sa/3.0/legalcode

a Launchpad account plus a Github account and use git for checking in and out docs.

We use the Maven plugin to automate builds on a Jenkins server,[14] but contributors can also build locally. The Maven plugin is developed and maintained by a Doc Tools team at Rackspace and dual-purposed for both Rackspace and OpenStack content output. It is available for anyone to use and contribute to on Github at github.com/rackspace/clouddocs-maven-plugin. For API documentation, our OpenStack tool chain is nearly identical to the Rackspace tool chain.

Here are some of the surprises I've had while working on OpenStack:

- Publishers want OpenStack content. I feel like I'm in a fight to be an acquisitions editor some days. Everyone wants an OpenStack book, or blog entries about OpenStack, or a magazine-style article to publish on their site. Sometimes I write the article or blog entry, but mostly I try to encourage other writers by reviewing their work, outlining for them, or simply embedding their blog entry into our documentation when it fits well.

- I once thought contributing to docs was a good entry point for new contributors. I now sense that it's really difficult for new people to contribute to the docs, and I also believe that developers should write for developers.

- Doc contributors need access to people they can interview incessantly. Plus they need access to hardware and a publicly-accessible cloud to try out their procedures. I couldn't solve this problem on my own. Two projects have filled this gap, TryStack, a place for trying out cloud, and Devstack, a set of scripts that lets you run your own integrated OpenStack cloud. We also use IRC often to ask questions.

---

[14] http://jenkins-ci.org

- I am shopping around the idea of holding an OpenStack tweet chat regularly. The idea here would be to try to reach non-IRC users – mostly developers who contribute to the project – who aren't contributing to the documentation yet or those who are consuming OpenStack as a cloud programmer or cloud operations pro.

- I have been amazed at the quality of content about OpenStack coming from bloggers. I am happily contacting them and incorporating Creative Commons licensed content into the official documentation collection.

- We are now transitioning to translation. Exciting times! I'm pleasantly surprised at the willingness of member companies and their international employees to take on a daunting task.

The future of our project depends on collaboration, and I am always looking for ways to bring in more collaborators and make it easier to work on the documentation. We also need to work on a "support knowledge base" as our Question and Answer systems are fragmented. There is plenty more work to do, but the project is already changing the world of cloud computing.

Gnome Foundation

Shaun McCance works as a Director at the GNOME Foundation. Shaun organized and hosted the excellent Open Help Conference in Cincinnati, Ohio, bringing together a group of open source documentation providers and community experts. We learned so much from each other in a short period of time. He has been working in this area for nearly ten years and has seen many of the models for open documentation systems.

Learning from GNOME: Interview

1. **How are you aligned in the company? Support, marketing, training? What is your job title?**

    GNOME isn't a company. Although it is backed by a 501(c)(3) non-profit organization, the foundation does not decide who works in which technical role.

    I am both a technical writer and a developer for GNOME, and I dabble in marketing. Most of my development work revolves around documentation: building, managing, validating, displaying, etc.

    I've been the documentation team lead since 2003. Semi-officially, my title is "Fearless Leader." My job is a mixture of writing, teaching, and cheerleading. I try very hard to write only as much as I need to lead by example and to spend my time teaching others how to write.

2. **What types of content do you deliver?**

    GNOME delivers topic-oriented help almost exclusively. We have some legacy manuals, but we've been steadily moving away from that format for the last three years.

    We deliver our help along with our software, and users can view it from within the help viewer without going to the web.

3. **How is your content licensed and how was the license selected?**

    GNOME is a strictly free software project, so our documentation must be licensed under a free content license.

    We license our content under the Creative Commons cc-by-sa 3.0 license. For developer documentation, we have a standard disclaimer that allows example code to be reused without restriction.

We used to use the GFDL (GNU Free Documentation License). The GFDL is very much designed for manuals: things that are updated infrequently and published when finished. We update continually and publish constantly. The requirements of the GFDL for displaying things like the history of changes were cumbersome. We began transitioning about two years ago.

We like copyleft licenses in GNOME. They ensure that what we do stays free for everybody always. Using the CC license was an easy choice because so many of our partner teams, like Fedora and Ubuntu, were switching to it as well. It makes it easier to share if we all use the same copyleft license.

4.   How many collaborators do you work with regularly?

At any given time, there's usually between four and eight active contributors. All of our contributors, including me, are volunteers.

We have a revolving door of documentation contributors in GNOME. I've worked on slowing that door, but I recognize that people will move on. People often use documentation as a stepping stone to contribute to other parts of GNOME.

5.   Do you hold sprints and if so, how many people are active in sprints?

We've held four sprints so far, with three of them being in the last year. We get six to ten people at a sprint. We've refined our sprint technique a lot, and we have a much better sense now of how to get the most out of them. In the future, we'd like to do two to three sprints per year.

6. **How many page views do your pages get?**

I don't really have any way of answering that question. I wish I did. Although we do build our help in HTML and put it on the web, we also ship the help with our software, so users can view it locally. This is the preferred form.

I can't get page view statistics without phoning home. I refuse to phone home without an explicit user opt-in. And I don't want to prompt users with an annoying opt-in dialog when they're already annoyed and don't want to spend any more time than they have to in the help.

7. **What is your tool chain and how is it created? Internally, with outsourcing, through the community?**

We write all of our help in source XML formats. We used to use DocBook for everything when we made manuals. Our new topic-oriented help uses Mallard. The source is stored in version control (git) along with our software.

We install our help along with our software. We actually install the source XML files. Our help viewer application is able to read these files and convert them to HTML on the fly for display.

We also build all our help to HTML and put it on the web for each release. The transformation code is based on the same code that's used in the help viewer.

GNOME is translated into over 80 languages, and we translate our help as well. Our translators are used to working with PO files, so we've built tools to extract messages from XML into POT files, and then merge the translated messages back to create localized help.

Almost all of our tools are developed internally, sometimes with the help of other open source documentation projects. I do most of the tool development work.

8. **Anything else you want to share about your particular situation?**

I think GNOME is doing interesting and unique things in open source documentation. Most other open source projects I know are doing either manuals or purely reference material. I think our team has made a lot of progress in figuring out how to do topic-oriented help well, and how to make it work within a volunteer community.

I'm personally bored with manuals. I just don't think you can do very interesting things when you're constrained with a linear format. Topics are a good first step, but I really want to see people exploring new and innovative ways of delivering help. People don't want to read the help. Rather than tell them to RTFM, I'd like to find new ways we can help users that they'll actually respond to.

Red Hat

Lana Brindley works at Red Hat and proudly dons her red fedora for talks. Red Hat is an open source technology solutions provider. I met Lana at the 2011 Open Help Conference and it has been great learning from her.

Learning from RedHat: Interview

1.   **How are you aligned in the company? Support, marketing, training? What is your job title?**

My job title is Senior Content Author. I work within the Engineering Content Services department, which looks after all our technical writers and translators.

2.   **What types of content does your team deliver?**

Primarily end-user documentation in the form of User Guides, Technical Reference books, and the like. However, we are also involved in producing Release Notes, Errata, and Technical Notes. We are sometimes called upon to provide editorial services for documentation produced by other departments, such as whitepapers produced by developers, customer information provided by marketing or sales, or internal documentation such as guides for internal processes and systems.

3.   **How is your content licensed and how was the license selected?**

Most of our documentation is distributed as CC-BY-SA, using the following licence text:

Creative Commons Attribution–Share Alike 3.0 Unported license ("CC-BY-SA"). An explanation of CC-BY-SA is available at http://creativecommons.org/licenses/-by-sa/3.0/. In accordance with CC-BY-SA, if you distribute this document or an adaptation of it, you must provide the URL for the original version. Red Hat, as the licensor of this document, waives the right to enforce, and agrees not to assert, Section 4d of CC-BY-SA to the fullest extent permitted by applicable law.

Prior to 2009, we published documentation using the Open Publication License. The Open Content Project deprecated the OPL and recommended a Creative Commons licence as a replacement around this time. We went with the suggestion, as CC had several other advantages. It worked better with the licences we used in our software, and allowed us more flexibility with using Red Hat documentation in the Fedora documentation projects. Also, CC has been more rigorously tested through the courts in many countries.

4. **How many collaborators do you work with regularly?**

This varies from project to project. Smaller projects are usually tackled by one writer, with assistance provided if needed to cope with a sudden large influx as work, such as a large errata coming in just before a release, or a large last-minute addition or change. Larger projects will have teams of up to about ten people, with one person project managing and ensuring targets are hit.

5. **Do you hold sprints (intense periods of collaborative activity for a deliverable) and if so, how many people are active in sprints?**

We don't hold anything as formal as a sprint, but we do have a corporate culture of everyone pitching in to help if a team is overloaded or might miss a deadline. We use a traffic light system in our weekly and monthly reports to indicate if help is required. Yellow indicates that a problem might be looming, and red is a call for help. People working on other projects that can spare time will volunteer whatever time they have (even if it's only a few hours) to help out a team in need.

6.   **If you have statistics, how many page views do your pages get or how many downloads for white papers and the like?**

I don't have access to this information.

7.   **What is your tool chain and how is it created? Internally, with outsourcing, through the community?**

Our toolchain has been developed in house, and most components of it have been made available in the open source community. As such, our tools continue to grow and improve with in-house development, but with testing and patches provided by the community. In its most basic form, we write in DocBook XML which is then turned into consumable HTML, PDF, and EPUB formats by our tool Publican, with version control managed by Subversion. We also have developed various other bolt-on bits and pieces that allow us to do topic-based authoring, non-linear publication, use different version management tools such as git, and other interesting things. What tools and formats we use depend largely on the individual project.

8.   **Anything else you want to share about your particular situation?**

Red Hat attempts to apply open source principles to everything we do. That means not only developing internal tools using open source development methodologies, but also ensuring that we allow the community to assist with growing and improving the documentation itself.

Open source starting points

In an interview with industry analyst Michael about open document-ation methods, I immediately jumped to "Who are you writing for?" as the very first question to ask. I think you also should ask, "What are they reading already?" Audience analysis is important every-where, but I would say it's even more important in open source be-cause much documentation effort is focused on the developer, which sometimes means non-technical end users get ignored. Also, there is so much free, liberated content in open source, you have to visit (and answer!) the question, "do we make the doc or gather it?"

I also said that FAQs are a perfectly good starting point, especially if customer support is your main goal. In an email exchange later, we talked about how documentation is a great conversion tool for website visitors. With web analytics, that measurement is possible. In essence, your documentation can be your storefront. Many people call FAQs "brain dead" and think they should die, but in the world of websites like Stack Overflow[16] we start to see the ultimate need for question and answer formats, especially when they go beyond simple forum designs into building portfolios and online reputations by answering questions.

I believe many open source projects rely too heavily on a wiki for their documentation toolset without also requiring a strong guiding hand. You also make gains by deeply embedding writers in the de-veloper community, and a wiki might not fit in well with existing developer processes. Also, researchers at McGill University conduc-ted an exploratory study about the documentation process of open source projects titled, Creating and Evolving Developer Document-ation: Understanding the Decisions of Open Source Contributors, available at www.cs.mcgill.ca/~martin/papers/fse2010.pdf. They interviewed 22 developers or technical writers who wrote or read the documentation and found that teams that started with wikis

---

[16] http://stackoverflow.com

moved to something more systematic. Their line is "We also found that all contributors who originally selected a public wiki to host their documentation eventually moved to a more controlled documentation infrastructure because of the high maintenance costs and the decrease of documentation authoritativeness."

Licensing considerations

Licensing is one way for content creators to indicate the terms under which they will give permission for their content to be used, sold, distributed, and so forth. Rights intended for legal copying of a work, also known as copyright, was intended to protect the creator from publishers publishing the content, "to the Ruin of them and their Families."[18]

Copyright, or rights surrounding copies, was a concept written into the US constitution. One famous American, Mark Twain, thought it was a good idea to extend his copyright privileges in 1906. Copyright law's international standardization started in 1886. Copyright law gives the original creator exclusive rights to do with the content as they please. A writer might give someone one-time printing rights for an article. A photographer might display a photo on Flickr but mark it with a certain license that indicates how you can use it.

The phrase "open source content" isn't precisely defined. The term open source typically describes software, but more and more often you can use the term "open source" to talk about content that can be reused as long as licensing requirements are met. You can see a discussion of various licenses used by documentation experts in the interviews included in this book.

---

[18] From the Statute of Anne, considered the origin of all copyright. http://www.copyrighthistory.com/anne.html

Documentation may lean towards openness, but requirements and restrictions on use, attributions, and changes (or branches) from the original document source vary widely. One strange example you can see on the web is a PDF that is licensed for reuse, but doesn't indicate any way to get the source so that you can reuse something other than the PDF itself.

Creative Commons has four licenses that are explained in straight-forward language on their website.[19] The Creative Commons website describes the three layers of their licenses – the human-readable layer, the machine-readable layer, and the legal code. This layered approach helps creators understand their rights, while also making the content findable on the web. The most accommodating Creative Commons license requires only attribution. Then there are more restrictive mixes and matches that address commercial use, whether you must share any derivatives, whether you can make a derivative, and so on.

Licensing of content is something that we've talked about at length on the FLOSS Manuals discuss email list.[20] As a counterpoint to those who tout the freedoms enabled by Creative Commons, Adam Hyde sees content licensing as limiting and cumbersome. He believes that even requiring attribution limits the "freedom" of content to go anywhere and be used by anyone. He truly wishes for source to be shared with everyone, everywhere, and creates tools that enable that vision. Collaborative content often requires that attribution of all collaborators be displayed somewhere, a requirement that can be difficult to meet.

Legally and ethically, documentation writers should understand the license under which the content can be used, and follow the licensing instructions. When selecting a licensing strategy, consider the readers, the creators, and the legal team's obligations.

---

[19] http://creativecommons.org/about/licenses/
[20] http://lists.flossmanuals.net/listinfo.cgi/discuss-flossmanuals.net

Not about tools

Because open source focuses on collaboration, tools that encourage collaboration serve the goals best. But this intent doesn't mean you have to use a wiki. In fact I learned an interest comparison between two open source documentation projects at the Open Help conference[21] in May 2011 in Cincinnati, Ohio. I have to talk about the dichotomy between using a wiki for community-contributed documentation and using an XML plus version-control system for collaborative authoring. Both methods were well-represented by Red Hat and Mozilla.

One of Red Hat's 60 technical writers, Lana Brindley, spoke about Red Hat's XML-based writing workflow in "Open Source Documentation in Four Easy Steps (and one slightly more difficult one)," a talk at the Open Help Conference in Cincinnati, Ohio in spring 2011. Red Hat has dedicated editors, a style guide, an IRC channel dedicated to grammar, and a search engine specifically created for writers to find content to reuse, plus a topic-based doc platform that allows writers to put together doc builds, which they built themselves when they realized they didn't want to build a component CMS. What she described was a very mature and high-quality, yet agile and flexible and open, documentation process.

Red Hat rivals any of the large enterprise documentation projects I've seen and accomplishes everything an enterprise needs to, yet with open source tooling, standards in XML, and somewhat hacked-together tool chains. Their content is translated to 23 languages, a feat only also accomplished by such companies as Dell, IBM, and Microsoft. Their culture sounds amazing, working for the "good guys" with tales of writers and developers working together to build the needed tool chains. I learned a lot about Publican and translations, about when to keep tweaking and when to release something to the wild.

---

[21] http://openhelpconference.com/

Community content strategist

After working as documentation coordinator on a popular open source project for more than a year now, I have seen a need for strategy on two fronts – the community front and the content front.

On the community strategy front, you want to build trust with your fellow contributors and encourage reputation building, a sense of collaboration towards a common goal, an eye towards the growth and acquisition of new community members, as well building a sense of belonging.

On the content strategy front, you want to devote the most energy towards care and maintenance and governance of the content that will meet the goals of the community, completing the circle of strategy. Being tactical about both of these strategies requires constant vigilance, constant analysis, but above all, a bravery in the face of the impossible.

We regularly manage about a dozen web properties that can be perceived as OpenStack documentation. Two of the twelve properties see about 100,000 visitors a month. Managing or coordinating efforts on this amount of content is a daunting task. Without hard-working dedicated people, smooth understandable processes, and robust well-maintained tools, you cannot meet the goals of your community. I believe a new job role is emerging as more of us bravely face the task of documentation in the open with fellow community members.

Concepts and Tools of the Social Web

As you begin learning about the social web or trying it out for yourself, you may become overwhelmed just by the vocabulary. This book offers a sampling of some of the terms that you might come across. While this chapter offers some definitions, you also need to realize that the meanings change, and often you will have to search the Internet or ask colleagues or friends for their definition before coming to your own conclusion about whether a new term is useful to keep in your vocabulary. Beware of trendy but meaningless terms, and be choosy about your definitions.

This chapter contains a frozen-in-time list of some terms and tools in 2009 – revisited in 2012 – that are related to social media, social networking, and social relevance. It isn't easy to define every term, and often the best way to learn a definition for one of these terms is to try the concept for yourself to get a sense of what it is all about. Or, ask someone who uses these tools often for his or her favorite use cases. Even those items that do not currently seem to fit your business goals may someday be applicable in another situation.

New media content categories

Table A.1 defines and describes some of the new types of content used on the social web.

Table A.1 – Types of Content

Category	Definition and description
Audio	Sound-producing audible file sharing, such as podcast sites, audio book sites, and music download sites.
Community content	Content that is central to a community. Wikis and forums are classic examples but also Facebook pages or Twitter hashtags can offer content that builds community. Polls, comments, and ratings are often used within communities for feedback.
Discussions	Content focused on talking, conversing, or discussing with people. Examples include blog or wiki comments, support forums, and email lists.
Email	Electronic messages sent between computers. Examples include Outlook, Thunderbird, GMail, and other mailers used to write messages electronically.
Instant Messaging	Real-time or near real-time conversation using messages. Increasingly, video and web cams are included in this category. Examples include Yahoo Instant Messenger, Skype, AOL Instant Messenger, IRC, and GTalk.

Category	Definition and description
Status updates and post-style messaging	Small posts of content often limited to a few characters. Examples include Twitter, Identi.ca, Chatter, and Tumblr.
Photo content	Websites and services used for sharing and organizing pictures. Examples include Flickr and Picasa.
Presence and location	Geospatial indicators tell where a user is geographically. Examples include mobile cell phone location "check-in" sites like Gowalla or Foursquare.
Profile content, contacts storage	Profiles that describe users, provide contact storage and lookup services, and give users an identity. Nearly all of these sites require registration. Examples include professional networking sites like LinkedIn and Plaxo.
Shared content	Content that people create to help others. Examples include blogs, wikis, photo sharing sites, video sharing sites, and bookmarking sites.
Syndicated content	Content that can be announced to interested readers using a notification means like RSS or email. Examples include blogs, podcasts, and news websites.
Tags and tag clouds	Tags are keywords assigned as metadata on content. Social bookmarking services like digg.com or deli-

Category	Definition and description
	cious.com use tags to help people retrieve content. Tag clouds are visual representations of tags that indicate relative popularity or the amount of content through typographical distinctions like the size of the text.
Text messaging	SMS (Short Messaging Service) is the standard for this type of communication, which is specific to mobile phone carriers. Text messaging is used to send short messages, photos, and links using a mobile phone.
Video	Sites used for sharing videos or screencasts and responding to video posts with videos. Examples include YouTube.com, Vimeo.com, Screencast.com, Instructables.com, as well as project or company-specific sites like Wordpress.tv and Sun.tv
Virtual	Used with avatars or other online objects representing a personality or a physical device. Examples include Second Life and haptic controllers like the Nintendo Wii.

Social web techniques

Tagging

folksonomy, social bookmarking, tag clouds, taxonomy

Tagging involves a skill with which many technical writers are familiar – determining which keywords best describe a blog entry, link, photo, video, or image for retrieval later. Social bookmarking sites like delicious.com and digg.com use tagging, or keywords, to help users find their links later or group them into categories.

What is more advanced and even social about tagging is that you can usually see the tags that others use, and also see the content that they've tagged. Tagging is a little more chaotic and less organized than more familiar techniques like indexing, so it might take some time to adjust your expectations.

For example, your company style guide might mandate that the gerund form of a verb is always required, or that plurals should be preferred keyword forms. But with tagging, any word or term is allowed, and sometimes the terms don't match – work and working are one example of such tag mismatch. The collective set of tags is known as a folksonomy – a taxonomy created by everyday folks. Where taxonomies are strict and rigid, a folksonomy is flexible and allows for any term. Both are powerful.

Tag clouds (see Figure A.1) are a visual method of displaying tags. Tag clouds use font size or weight to indicate the amount of content or popularity of each tag. You can use websites like wordle.net and tagxedo.com to build these word clouds yourself.

Figure A.1. Visual cloud of words created on Wordle.net[3]

While social bookmarking and social tagging are some of the least adopted technologies on the social web, these very same technologies also show the most promise for businesses as they share information and collaborate internally and externally with others.

According to a Gilbane Group report, *Collaboration and Social Media—2008*[47], although only 5% of those surveyed said they currently use social bookmarking, nearly half of those using it rated it very effective.

Customer support groups can share how-to and troubleshooting information by using tags for links that answer customer questions. One technical publications group collects the tagged links from the support group once a week as a report on the most asked and answered questions.

---

[3] Technically this is not a tag cloud since the words were collected from the feed for my blog, justwriteclick.com, but it illustrates the concept.

Tagging examples

In 2003, the website del.icio.us, now renamed delicious.com, offered users the ability to apply metadata as tags on stored bookmarks for easier retrieval later. Users also share their tagged bookmarks with others on the site.

Pronounced "flicker," the photo-sharing site Flickr has an explore-by-tags feature (see Figure A.2) that shows the popularity of tags like "birthday" and "family." The Hot tags area shows the most common tags used for recent uploads, and the font size of each tag indicates its popularity on the site.

Figure A.2. flickr.com

Shared content

community content, user-based content, participatory media

If you study social media, you'll see that production and distribution changes are at the heart of these emerging methods. Naturally, one concept or skill that technical writers know thoroughly creating content. What is interesting about the recent mind shift towards community content is the thought that "everyone's a technical writer."

In reality, of course, everyone is not a paid, professional technical writer, and hopefully paid writers have skills, experience, and practice that set them apart from the rest of the crowd. However, where user-based content lives, there is a passion for the product, and customer forums and fan-built web sites display that passion for all to see. Consumer-generated media, as it is also called, can refer to many different kinds of media, not just written text, but video, audio, and images.

A conversation can begin with a user creating a rough draft of content, even as a question on a support forum. A technical communicator then builds a mental concept based on the user's perspective through their contribution. Comments on wiki pages and two-way conversations on online forums are a prominent part of this type of content sharing, mixing, and modifying.

Publishing models get turned over when anyone can publish content and make it instantly available through Internet channels. A few years ago it wouldn't have been possible to view published content as easily and quickly as we can now. Of course people argue that the more content there is, the less peer-review (or any review or editing) it gets, so therefore it must be of lesser quality.

That assumption is not always true, especially when you consider whether the content solves a problem quickly. A higher search ranking may prove that the consumer-created content is in fact helping more people. On the other hand, if someone publishes an incorrect message or workaround, perhaps you reach out to see how you can correct the content or that person's perception.

If you read through comments from your users, you may find sophisticated readers who have high expectations for the consistency, organization, and find-ability of your documentation.

You may also find that every representation of your company is thought of as "technical documentation," including white papers or data sheets on your website that were not created in your department. I have been pleasantly surprised by suggestions from a customer that accurately pinpointed an area to improve based on the latest best practices and standards for online documentation. I'm certain that today's sophisticated information seekers have high expectations and concrete suggestions.

What also might follow, as your customers verbalize their hopes for product documentation, is a willingness to assist. So, another interesting aspect of this new publishing channel is what motivates people to contribute content. When it is your job to create the content, like it is for technical communicators, money is your prime, though not sole, motivation. Other content creators may be motivated by wanting to be known as an expert in the field, wanting to help others so that they'll be helped in the future when needed, wanting to be part of a group or a greater cause, or finally, wanting to save time by publishing something once instead of repeatedly sending the same message.

Shared content examples

The World of Warcraft wiki[6] is a popular wiki for players of the massively multiplayer online role-playing game (MMORPG). Recently ranked in the top 800 websites in the United States by Alexa, it contains articles about the game's characters, game play guides, customization guides, and strategy information. Figure A.3 shows an article about combat that links to armor and weapon articles and describes concepts of combat in the game.

---

[6] http://wowwiki.com

Figure A.3. WoWWiki

With the tagline "The How-to Manual that You Can Edit," wiki-how.com (see Figure A.4) contains how-to articles with topics ranging from hobbies to holidays. In January 2009, the site surpassed the 50,000 article mark[89].

Figure A.4. wikiHow

Syndicated content

podcast, blog, web log, vlog, video log, RSS, subscription, feeds, aggregation, news

All of these terms are related to content that is updated regularly. Blogs are typically text content that is updated often; sometimes additions are made multiple times a day. Blog entries range from a few sentences that make a comment link to another site to fully developed essays. Blogs can be personal, professional, or anywhere in between. You can find the original definition of web blog and why it became blog on Jason Kottke's blog.[8]

Podcasts are audio files that are posted regularly. Often podcasts are posted using blogging software, so there may be overlap between the two types. Podcasts are much like a radio show featuring interviews.

A vlog or video log has regularly posted video content. An example is rocketboom.com, which was a news show featuring an amateur newscaster who became so popular she often had more viewers than the local news affiliates in her area.

So how does regularly updated web content compare to your online help experience for your users? Have you already implemented an online help system with dynamic updates, or are you still tied to the software release schedule? As customers become more accustomed to the immediacy of blog updates, they may increasingly expect frequent updates.

Also, consider writing a blog yourself, from the viewpoint of a dedicated technical writer, to increase awareness of your customers that you are a real person inside the company, working hard at providing accurate, relevant information. You can be part of the customer conversation that blogging provides.

---

[8] http://kottke.org

While social media metrics are still in their infancy, many studies show that the reach and influence of bloggers can be and will be measured. For example, blogging about information development or architecture issues that you wrestle with in the name of high-quality documentation might become a good selling point for your company's product, giving the customer confidence that you are giving your best towards helping them solve business problems.

One example of a blog that gave the technical writer a voice in the corporate world is my former blog on talk.bmc.com. The salespeople would get very excited if a potential customer reading the conversations on the blog led to a one-on-one contact. White paper downloads are tracked through the links from the blog authors as well, and blog readers are also likely white paper consumers, doing their homework with searches. Industry analysts also read the blog, and got to know some of the bloggers, which in turn helps the analysts understand the goals of the business and interpret the values and strategies that the company employs to serve customer needs.

Another trend is the use of blogs for describing new features in products, especially web applications. Examples include a blog entry describing new Google Calendar features[11] and SmugMug's blog[12] where the entire blog is dedicated to release information.

The Jing product's online help[13] is written and maintained in Movable Type, a blogging tool. Jing is a relatively new screencasting tool. Many blogging tools can be used as content management systems, and it appears that Jing's writers see blog engines that way, too. The writers are taking advantage of some nice built-in features, such as a Search field at the top of every page, and the Categories link at the bottom of each help topic give an automated collection of topics. They have just one "table of contents" for the help system,

---

[11] http://gmailblog.blogspot.com/2009/01/get-calendar-on-your-google-desktop.html

[12] http://blogs.smugmug.com/release-notes/

[13] http://help.jingproject.com/

and that's the top page, but it works nicely as a site map. The overall effect is a very simple and elegant user assistance or support system.

A blog can work well for release notes. Give your product some "Google juice" – the mysterious quality that raises the ranking of a page in Google's search – and generate buzz for new features by giving other bloggers a well-understood infrastructure to link to you and give your entries trackbacks.

This model works best when your release notes are not the primary vehicle for reporting bug fixes, since the open, honest style of blogging may reveal more than your marketing department would like. Separating feature information from bug reporting may help you "spin" features without guilt, while keeping the information factual.

Syndication and RSS

The *RSS* acronym stands for Really Simple Syndication. RSS and *Atom*, a similar standard, are XML-based formats for distributing and gathering content from sources across the Web, including newspapers, magazines, wikis, and blogs. This syndication ability is the real beauty of blogs and podcasts. You write a blog entry, upload a podcast, or edit a wiki page and people get notified when that new content is posted. According to Technorati, blog subscriptions were doubling every five months in 2006[85]. Subscription technology is a big deal.

RSS, one of the few three-letter acronyms in blogging and podcasting, is not all that technical. RSS and feeds are nothing more than notifications and alerts for new items, be they new blog posts, new or changed wiki pages, or even package tracking notifications.

RSS aggregators are tools that manage a collection of subscriptions for you. You can find popular RSS aggregators or readers on the web (see http://blogspace.com/rss/readers). The first aggregators for reading blogs were standalone readers (Sharpreader is an example) and then web-based readers were created. Bloglines and

Google Reader are popular web-based readers, and some browsers have this ability built-in. Feed reader tools often offer a combination of notifications via email, mobile devices, or on the web.

For audio content, like podcasts, people can load an audio file to a player such as an iPod, listen using an application such as iTunes on their computer, or just read about the podcast before deciding whether to download the audio or not. How people want to listen to content dictates which feed they choose.

Feedburner is a Google-owned tool that offers subscription pages that contain buttons for the most popular aggregators. All you have to do is click one of the buttons to subscribe to a feed with that particular aggregator. Figure A.5 shows examples of the list.

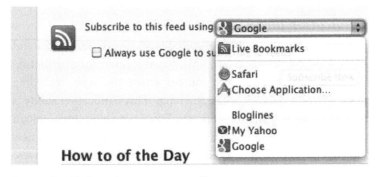

Figure A.5. Subscriber aggregator list

On some sites, feeds are available that allow you to track subscriptions and gather detailed statistics like what browser a reader uses to access the site, whether it's a robot that's crawling the site, and so on.

For podcasts, sometimes there are different icons that represent different audio feeds. For example, you can subscribe to a feed that downloads the audio files directly to an iPod or other audio player. Another RSS feed downloads the audio files directly to a computer and plays them in iTunes. Yet another feed offered by some podcasters gives text-only summaries of the podcast containing notes for each podcast. People subscribe to the summary feed when they want to know about the podcast topic and contents before downloading the audio file.

If you click an RSS icon, your browser will normally use a default feed manager. However, sometimes the website has linked the icon to another page. For example, you might be redirected to a page that contains aggregator subscribe buttons. If you right-click the RSS icon, you can copy and paste the feed URL into your aggregator.

I am amazed by the ways people use RSS for notifications. At a South By SouthWest Interactive (SXSWi) session I attended in 2006, the panelists presented a list of sample ideas you can do with RSS and a feed reader such as Bloglines or Google Reader.

- Get new/updated podcasts or videos by subscribing to a feed on iTunes or Apple TV. RSS subscriptions are not just for text notifications.

- Find the most popular blog entries and topics with RSS – aggregation of information is crucial for this goal. You can enter a product name and continuously get information when a new item is created with that keyword.

- Find time-critical information or immerse yourself in a point-in-time. On some feed readers you can search within a specific date range as well. Bloglines has added a longer time span to their return content feature, allowing you to read feeds from 2001.

- Get updates to classified ads based on a search or a category. For example, if you are seeking a specific pre-owned electronic device for a cheaper price on Craigslist, you can search for that keyword within the category, then copy the RSS feed link and put it into your feed reader. Any time a new item is added to the craigslist classified ads, the notification appears in your feed reader.

- Follow group conversations by RSS instead of using email and clogging up your inbox with multiple messages. Yahoo Groups offers RSS feeds for conversations, so you can read them from one place such as Google Reader.

- Track packages using RSS with a website like http://simpletracker.com. Some eBay resellers say that subscription-based tracking is better than using repeated email messages for notification.

- Use RSS subscriptions as an email subscription alternative for product newsletters. It still requires customers to opt-in, but it lets them choose RSS instead of email.

- Use an RSS calendar to turn calendar dates into RSS notifications for birthdays, anniversaries, and so on. The website is http://rsscalendar.com/.

- Give notification for course information via an RSS feed.

- Create keyword or hashtag searches with Twitter's search engine[14] and subscribe to RSS notification for competitive intelligence and other specialized information, and receive notifications only when the Tweet is in your favored language. Language filtering is especially useful for a product name like the one I've worked with, iMIS, which is a Spanish word that tends to show up in Spanish language posts. Restricting the results to English-only posts eliminates the non-relevant Spanish content.

- Search for job leads on indeed.com with RSS notification; you subscribe to search terms, and search for your product to view the job descriptions of your typical end-user. For example, search for Remedy in Arkansas and 29 job descriptions show up, but if I subscribe to the RSS feed for that search, I can be notified when each new job listing appears.

---

[14] http://twitter.com/search

Think of how a subscription-based notification on end-user documentation changes could help your customers. Would a subscription to release notes additions help your customers? You do need your customers to want the constant updates, but since they control the amount of notifications by managing their subscriptions, they can choose the interval and amount of updates they receive.

Syndication examples

Blogs written by writers, such as this blog (http://blogs.adobe.com/-indesigndocs, see Figure A.6) by the lead writer for InDesign and InCopy at Adobe, Bob Bringhurst, can give writers the opportunity to provide additional explanation about a feature. A subscription to this blog gives people information, updates to the help, and general tips.

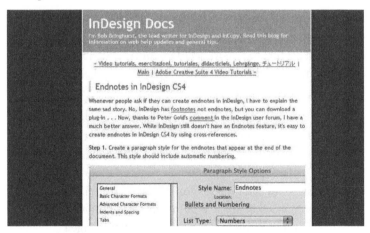

Figure A.6. InDesign Docs Blog

By subscribing to updates, readers of this Embarcadero blog by the technical publications manager Dee Elling can find out when a new set of documentation is released, such as the PDF files that this writer announces in this blog post. She also asks the readers about their favorite formats and invites feedback through her blog.

Figure A.7. Embarcadero blog – Dee Elling

Community

wiki, user-based content, crowdsourcing, discussion

Community-based content can be a collaborative effort on a website, or an email list that has a core set of members contributing information that furthers the community's goals. As Rachel Happe states in her blog entry "Social Media is not Community"[58]:

> Communities gather around a concept or common goal, not around a collection of content (although content does plays a major role, it is not the impetus for the community). So while wikis are a type of community-based content, a community cannot be interchanged with the social media tool itself that enables working on the community goals.

Wikis are a technology for creating web pages quickly. They allow anyone with the correct permissions to edit any page on the wiki. Created by Ward Cunningham in 1995 to make web pages for a consulting firm, wikis work across multiple platforms and browsers and encourage collaboration.

Many wiki applications have sophisticated methods for resolving simultaneous edits. They encourage what is called "crowdsourcing," organizing an open call for contributions from anyone with interest and ability. *Crowdsourcing* was a term used in a 2006 Wired article by Jeff Howe[62], but the concept dates back to the 1800s when the Oxford English Dictionary was created out of slips of paper with well-researched definitions submitted by volunteers.

For technical publications, wikis are an exciting possibility for engaging the customer and getting the customer's viewpoint on choices and tasks for concepts as well as scenarios for how they are using the product. Wikis offer living, breathing documentation because users can change pages based on new information or the results of a discussion about the content.

Readers have become familiar with wikis and may bring certain expectations to them. Some readers may consider any collaborative site to be a wiki. Customers might expect a wiki to be in place for certain products, such as open source technology or games. Even if your product doesn't seem to support a group that would demand transparency and openness in the documentation, think about ways to offer collaborative tools, even if you don't offer a wiki.

I found the OLPC wiki at wiki.laptop.org very useful for project planning and recruiting. When I needed artwork to demonstrate how to open the OLPC laptop's unique case, I posted a request on the Artwork Wanted page. I had two excellent figures sent via my wiki "Talk" page within three weeks. Also, I could talk to other wiki users by using their wiki "Talk" pages rather than relying on email.

Because OpenStack, our wiki at wiki.openstack.org, does not enable comments, one of the major use cases for a wiki is stopped at the start. Our project planning and project documentation work occurs on the wiki as well as detailed software specification authoring. Most of the user-documentation oriented pages typically point to another site for how-to, conceptual, or reference information, where comments are enabled. One odd observation is that people expect our developer documentation, which is built from Restructured Text using Sphinx, to be a wiki when in fact it is not.

Wiki examples

- **Adobe Labs:** http://labs.adobe.com/wiki/index.php/Main_Page
- **Ajax patterns:** http://ajaxpatterns.org
- **Apache wiki:** http://wiki.apache.org
- **Confluence wiki:** http://confluence.atlassian.com
- **Knoppix wiki:** http://www.knoppix.net/wiki/Main_Page
- **Microformats wiki:** http://microformats.org/wiki/Main_Page
- **Motorola Q:** http://www.motoqwiki.com
- **Mozillazine Knowledge Base:** http://kb.mozillazine.org/Knowledge_Base
- **One Laptop per Child wiki:** http://wiki.laptop.org
- **Opera Browser and Internet Suite:** http://operawiki.info/Opera
- **SugarCRM Wiki:** http://www.sugarcrm.com/wiki/index
- **SQL Lite:** http://www.sqlite.org/cvstrac/wiki
- **SugarLabs wiki:** http://wiki.sugarlabs.org
- **Ubuntu documentation:** http://help.ubuntu.com/community

Presence and location

BrightKite, Latitude, Twinkle, Glancee, geo-spatial location awareness, poke, status

You may find it odd that there's a function on web applications like Facebook to "poke" another user. These mechanisms are meant to get another user's attention or to find out if they are online or not, which that user can indicate by poking you back.

Some web applications integrate with GPS hardware in mobile devices to identify the location of users. This data is called geospatial data, and content systems that know the reader's location can help you analyze the localization or translation of content. For example, if you notice an upsurge in questions about a particular help topic coming from the Ottawa, Canada area, you could look for French-Canadian culture issues with the content and fix the issue. You can also find nearby people when you are at an event using an app like Highlight from Glancee.

TripIt and Dopplr are examples of travel web applications where you share your travels with friends and coworkers in the hopes of keeping others informed about where you are so that in-person meetings can happen. Mobility and status knowledge might be useful if you attend your user's conferences and want to set up face-to-face gatherings.

Presence example

By enabling more real-world interaction with a web application that collects location data and makes it meaningful to people, Dopplr (sounds like "doppler") gives you trip tracking and trip planning abilities. Figure A.8 shows a map of my trips over a year's time. It integrates with popular social networking sites like Facebook and LinkedIn to tell my network about my current location, and indicate upcoming trips that I have planned.

Figure A.8. Dopplr (http://dopplr.com)

Microblogging and post-style messaging

Twitter, Identi.ca, Jaiku, Tumblr

Twitter, or other microcontent sites, can also offer customers updates on the status of your work. Twitter is a web-based application that gives you 140 characters to answer the question, "what are you doing?" The now-closed Pownce website offered mobile integration so that cell phone users could triangulate position if the user chose to reveal it to their friends. Facebook and LinkedIn have a status application built into the Profile page, where you fill in an answer to "What's on your mind?" or "What are you working on?" or other fill-in-the-blank completion statements for your current status, state of mind, or whereabouts.

Twitter's character limit can be a difficult adjustment as you figure out what you can communicate in such a small amount of text. However, integrating IM-like or SMS-based communications into technical documentation might provide an innovative delivery method for instant online help when a customer needs it. Think of the Tips and Tricks documentation sometimes offered as a beginning blurb when a software product launches. Twitter's "tweets" are a similar small package of information.

Provide a Twitter feed of tips and tricks for your software, or news updates when a critical wiki page has been added to, or find ways to have IM-like discussions with your services or consulting teams while they are out on customer service calls. If you are familiar with the XML-based Darwin Information Typing Architecture (DITA), you could also think of Twitter posts like the `<shortdesc>` element required in every DITA topic that summarizes the topic.

You could also use microblog posts to offer a tip of the day and link to an online help topic or support article each day. You can use Twitter helper applications like ping.fm to schedule posts to many social networks. Sarah Maddox recently posted her experience with using Twitter to announce changes to her company's software product with links to release notes. She described their use of hashtags and offers a detailed write up in this blog entry, "Twitter as a medium for release notes"[68].

With Twitter, you can also put keywords, or hashtags, into your posts. This provides a way to categorize your messages and to follow a set of messages on the same topic. Twitter is especially useful if you are at a user conference and want to communicate with like-minded individuals. For example, at the 2008 STC Summit, Twitter users could enter the hashtag #stc08 anywhere in their Twitter post, and everyone could view those posts at the now-defunct site, twemes.com. In 2009, the crowd chose #stc09 as the hashtag for Twitter posts related to the 2009 STC Summit, and subsequent conferences have followed the same pattern. By 2009, Twitter's search page (http://twitter.com/search) was available, making it even easier to find hashtags related to the conference.

> Beware of relying on Twitter servers for long term storage – tweets are only saved for a limited period of time. The amount of time varies, but it can be as short as one week.

One of my favorite stories about Twitter's usefulness at real-time events is from SXSWi 2008. At Tom Parish's social media metrics panel, I sat next to Summer Huggins, who worked in Austin for Hammock, a media company based out of Nashville. We chatted about Austin, how I feel like a tourist in my own town when I come downtown, and laughed about the Compass Bank building when she said it looks like a giant nose hair trimmer. (Yes, it does.)

When she left the talk, she accidentally left her digital camera on her seat. I noticed it and asked both of the attendees in front of me if they caught her last name so that I could try to find an email address for her in the SXSWi attendee directory. None of us could remember her last name.

So, I took the camera in its cute case to the SXSW information desk to be placed in Lost & Found, telling them that someone named Summer from Austin would hopefully pick it up. They said they'd be open until 8:00, and it would be kept in a safe place overnight.

Later that night, I started searching on Twitter for someone named Summer from Austin who maybe, just maybe, had talked about attending the social media metrics panel. Sure enough, @SummerH posted a tweet marked with #SXSW[19] saying she had lost her digital camera! I immediately sent her a direct message on Twitter, clicked through to her blog, found her email address, and sent her an email telling her she could pick up her camera at SXSW Lost & Found. Problem solved!

Summer was very excited and also noted the power of the Twitter and SXSW attendees by tweeting, "Lost camera is in-hand. How much do SXSWers rock?!?!"(Figure A.9).

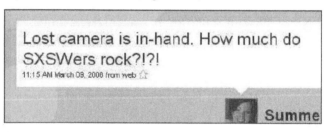

Figure A.9. A Twitter status update

Twitter can also be useful for crowdsourcing answers to simple questions. One example is the reaction David Pogue's followers had to his demonstration of Twitter by asking "I need a cure for hiccups… RIGHT NOW! Help?"[20] Unfortunately, he didn't indicate that he was merely demonstrating the power of Twitter for a confer-

---

[19] http://twitter.com/SummerH/statuses/768730904

[20] http://www.nytimes.com/2009/01/29/technology/personaltech/ 29pogue-email.html

ence in Las Vegas. He had responses nearly immediately but after those who answered his query found out he was showing off Twitter to a crowd of 1,000 people, their reactions ranged from humor to annoyance to feeling used.

I had an experience where I mentioned on Twitter that I was looking for Super Mario Jibbitz for Crocs shoes. I had two responses pointing me to sources for the item by the end of the day, from two different parts of the world.

Helpful Twitter users can make a difference in the corporate world. TurboTax has customer support representatives with a Twitter account at twitter.com/turbotax. The Twitter bio section for that account gives the names of real people. They post hints, but they also talk to people who have questions. If your technical publications role is closely aligned with the technical support department, you might consider starting one with tips, tricks, and helpful information.

You can enhance a blog with a Twitter account. I was surprised when my number of Twitter followers passed my number of blog subscribers in 2009. I now link to blog entries from my Twitter account. I was originally reluctant to do so (in 2007) because I thought people who followed me on Twitter would be reading my blog from somewhere else, like a feed reader. In 2009, I tried using a WordPress plugin to automate the twitter post for blog entries, but later abandoned the idea because I wanted to generate discussion, not just get people to click-through to my blog.

Microblogging example

The Twitter interface (see Figure A.10) is as simple as the Google search engine form. It contains a single field with the label "What are you doing?" You enter up to 140 characters in that field, click the update button, and your microblog post is sent to all your followers. By following people on Twitter, you create your own "timeline" of their micropost updates.

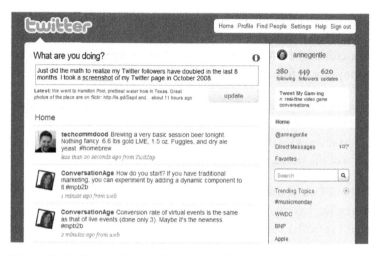

Figure A.10. Twitter (http://Twitter.com)

Profile

LinkedIn, Naymz, Spock, MySpace, Facebook

Profile sites let users enter information about themselves (or others in the case of Spock, a people search engine). This information is searchable and can be correlated by entering information about relationships to others as well as your activities online and offline. Human Resources recruiters have found this source of information to be quite valuable, and sales and marketing, branding, and research are also aided by studying these sites.

Much of the excitement and media coverage surrounding social media occurs when a large site that collects profiles about their users changes the way that data is displayed. For example, when MySpace began showing more distant relationships and activity that reached farther than users expected, users demanded that the feature be removed. Even though the data collection made it possible to see the relationships, users did not necessarily want these extended relationships revealed. Be aware that profiling can be misused, and privacy and portability of an online identity are controversial subjects.

Profile examples

LinkedIn (see Figure A.11) is a networking site used primarily for keeping up professional contacts, performing job searches, writing recommendations for others, reading company profiles, and storing contact information for co-workers both current and previous. It is very useful for finding information about end-users of the products you document, such as their job titles and where they work.

Figure A.11. LinkedIn (http://Linkedin.com)

Facebook (see Figure A.12), which originally allowed only college students to register on the site, has become hugely popular. According to Wikipedia,[22] Facebook had over 800 million active participants as of July, 2011. Of Americans who were online, Facebook's visitors averaged the most time on the site per person (7 hours, 9 minutes) according to Nielson Online,[23] you should observe Facebook for trends and usage patterns if your customers are Americans. You can learn more about your specific demographics with some research.

---

[22] http://en.wikipedia.org/wiki/Facebook
[23] http://blog.nielsen.com/nielsenwire/online_mobile/march-2012-top-us-online-brands/

Figure A.12. Facebook (http://Facebook.com)

Discussions

forum, board, email list, mailing list, instant messaging

For technical writers who are already affiliated with customer support, the idea of joining customer support forums or moderating and joining in conversations there is not a new one. Valuable relationships and much needed information are abundant in these online communities. Communicating on forums or boards is an excellent way to establish yourself as part of the user community.

Instant messages

Online instant messenger services such as Yahoo, MSN, AOL IM, Skype, or other IM clients enable you to be available to customers directly. Instant messaging clients sometimes include video and screen-sharing capabilities which is extremely useful for collaboration and explanation.

In her blog post, "Agile Tech Writer"[67], Sarah Maddox talks about answering Instant Messages from all around the world and being available to others to assist or just share humorous stories. Just be sure to have a sense of when to be unavailable, or keep "office hours" so the instant messaging communication doesn't become too disruptive to your work.

Another real-time chat technology that has been around for many years is Internet Relay Chat, or IRC. Writers working on open source projects find the instant chat capability valuable for both contacting developers around the world and getting support when needed. IRC has many different clients that enable chatting and also web-based methods for joining "rooms" or "channels."

Online forums and mailing lists

In a forum, people can post questions about a product in hopes of getting answers quickly. Many open source products use forums as their sole support mechanism. Forums allow write posts or respond to other's posts. More capable forum software allows users to include HTML and emoticons in the text, and include signatures. They also typically have fully-featured notification and moderation systems.

Community features include rating systems for each post or reply, avatars or profiles representing an online identity, private messaging, and even an indication of which users are currently online so people can talk with one another. You can use forummatrix.org[24] to compare the features available in different forum software offerings.

Many forums use point systems or other methods for earning ranks, company t-shirts, or even more valuable rewards, like registration to a user conference. In some systems, you can vote on the helpfulness of an answer or thank another forum member for helping you. Cory Doctorow's science fiction book, *Down and Out in the Magic Kingdom*[7], contains a concept called "whuffie" – a reputation-based currency that emerges after all of society's other needs are met for free.

---

[24] http://www.forummatrix.org

Tara Hunt has a book called *The Whuffie Factor*[14], about using the idea of whuffie to build online communities for business improvement.

There are others building systems for online community reputation building. For example, Yahoo Developer Network[25] contains a library of design patterns for reputation solutions. forums are an accessible way to view and measure whuffie in an online system. In Figure A.13, compare Guru status to Enthusiast – Guru has more stars, likely earned with "whuffie" equivalent on the forum, such as the number of times the author has been thanked per post.

Figure A.13. Online reputation system

Online meetings

Online meeting tools make formal presentations and meetings easier to host and facilitate. Two-way or multiple participant conversations can be held with voice, chat, presentation slides, and white boards shared with participants. WebEx, GoToMeeting, Campfire, and Dim-dim are just a few examples.

Mashups

Mashups, proximity data, multimedia

A mashup is a recombination of two or more sources of data to create a new deliverable or service. A classic example of a mashup is a map mashup, in which map data is overlaid with other data, such as favorite restaurant locations. Or data from event planning software that shows you which conference centers are booked for which dates, overlaid on a hybrid street and satellite image map that you can zoom in or move around. Google+ has a "Nearby" feature that gives you a feed of posts from people near you based on geolocation data. The feature works best with a mobile phone where geolocation is built-in, but you can also use it with a browser by adding a geolocator function.

Part of the idea of social media is that data can be combined in new ways. In 2009, Michael Priestly demonstrated how different DITA-based feeds can be recombined in different deliverables with the IBM Custom Content Assembler. This work has been extended and will be incorporated into the IBM Knowledge Center, which is designed to replace over 700 separate Eclipse-based infocenters at IBM. At the time of publication, this project was headed into its second public pilot.

DITA is an XML-based architecture for authoring, producing, and delivering technical information. Users select and organize (reorder) the topics that they want in a custom "map" and the user-created compilation is both searchable and reusable.

A similar capability is available in the Remix area of the FLOSS Manuals website.[26] To make a manual remix, you select a group of topics and then drag and drop them into the "remix box." You can then export to PDF or HTML.

---

[26] http://en.flossmanuals.net/index.php?plugin=remix

Several years ago Microsoft introduced a powerful format and data display mashup in their online user assistance. When a user selects certain links in the online help, instead of linking to a description of the task, the link opens a dialog box that lets the user complete the task directly.

Mashup example

The FLOSS Manuals remix I built for the One Laptop per Child project enabled us to write separate manuals for the hardware and software, plus we could deliver HTML help that shipped with several thousand laptops around the world. The remix is also used to export to "widgets" that can be embedded in HTML pages to populate content on other websites outside of FLOSS Manuals. See Figure A.14 for an example of the FLOSS Manuals remix interface.

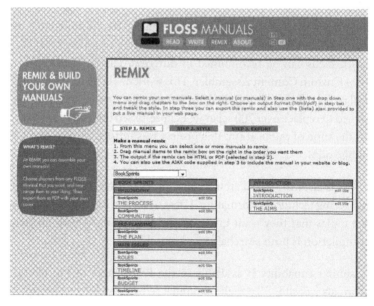

Figure A.14. FLOSS Manuals REMIX feature

Photos, video, media

photo sharing, video sharing, viral video

For products that need visual explanations, you can post photos to Flickr or videos to YouTube to demonstrate tasks or time-saving techniques. If you use tagging on Flickr, users can contribute screenshots with annotations about how they've used or customized your product. For example a user might post an iPhone screenshot showing a customized row of icons, along with embedded how-to information.

For a marketer, a video "going viral" is a dream come true. Going viral means that a video, image, or other communication becomes widely distributed, far beyond its original audience. Content may go viral because of exposure on a popular blog like Boing Boing,[27] or by word of mouth.

One example of going viral is the Subservient Chicken website, which let people play a flash animation to "have it your way." The story is chronicled on Wikipedia.[28]

I don't believe going viral is a necessary goal for end-user document-ation or for building a community around a wiki. But from a tech-nical marketing perspective, it would be great to have a link to a white paper or a tip of the day go viral on Twitter through RTs. Twitter posts can reach many people through the "retweet" feature where people send a copy of a twitter post along with the label "RT" and an attribution to the original poster.

If you are fortunate enough to create a viral link or video, be sure it's accurate and helpful to your users before it gets 12 million views on YouTube. Due to the speed of forwarding, "going viral" can be as dangerous as it is useful.

---

[27] http://boingboing.net
[28] https://en.wikipedia.org/wiki/The_Subservient_Chicken

Media example

YouTube is a video-sharing site that converts uploaded videos of any format into Flash-based or HTML5-based videos that are easily played back in a browser. People can also embed videos from You-Tube onto other websites. Figure A.15 shows a screen capture of a YouTube video that contains instructions for installing a bulk ink system for a printer. By creating a video response or text comment to a video, YouTube users "talk back" to each other and spark discussions about the video.

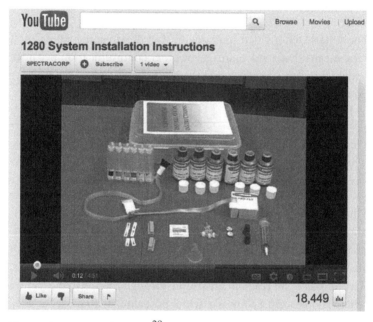

Figure A.15. youtube.com[29]

---

[29] http://www.youtube.com/watch?v=6c_a565qXxg

Virtual content

virtual reality, Second Life

Training organizations are already finding the value in offering training programs on Second Life, where people can gather with avatars that represent their personalities online. Second Life is a virtual three-dimensional environment where users create the buildings, land, and entire environment as well as clothing for avatars that represent them in the virtual community. Once in Second Life, users connect, socialize, chat, and interact with each other in the visual three-dimensional world.

For products that need a visual demonstration, there may be value in developing a training area in Second Life, but you need to consider development costs.[30] Also, consider the maintenance costs of virtual training environments, and whether the expertise will be available for future modifications to the training curriculum or coursework.

Figure A.16. Second Life auditorium-style meeting room[31]

---

[30] You may also find that the risk of "griefers" tearing down your virtual training environment is greater than the efforts spent in creating the environment.

[31] http://www.flickr.com/photos/krossbow/2717822171/

Virtual world examples

Second Life contains learning areas such as the auditorium in Figure A.16 used by Discovery Education. The virtual screen on display is called a Smartboard.

The three-dimensional nature of the world is shown in Figure A.17 at a seated gathering of students. People design avatars to represent themselves in the virtual world, and you can see the body language indicated by the avatar leaning over to talk to an avatar in another chair in the figure.

Figure A.17. Second Life informal graduate student meeting[32]

Learning about social media

Finding a definition for a word like "wiki" can be difficult due to the proliferation of occurrences in web pages that may not be relev-

---

[32] http://www.flickr.com/photos/pathfinderlinden/157712296/

ant to the topic of wikis. From Cookiepedia to Wookiepedia, there are so many wikis in use that it can be difficult to discover a definition that suits your needs. One method for searching for a definition of an unfamiliar term is to use the Google **define:** command. For example, in the Google search box type "define: wiki" to get a list of web pages that offer definitions for the word "wiki."

You will learn more about social media and tools for the social web by trying them yourself. Even if your company does not have a need to use these tools yet, try them for personal use or as part of a volunteer effort for a non-profit cause.

Evaluating newcomers to the scene

Whether you started with LinkedIn or Facebook, you probably started receiving Naymz, Spock, and Google+ invitations that spread like wild fire. How can you evaluate these newcomers and determine which ones have any return on the time spent with them?

Here are some ideas for evaluating newly available tools:

- Look carefully when you experiment to find out where your users have their conversations. The strength of a community is more important than the tools used to reach that community.

- Determine which types of people, such as small business owners or representatives of a corporate brand, are using certain tools and ask them what gains or returns they are seeing.

- Figure out your goals for social networking. This will help you figure out if you need to be on every single network or if a specific few that are directly related to your goals are the ones to focus on (most likely, it will be the latter).

- Ask a few people for their opinions on a tool – if they give it a thumbs up, then try it for yourself. This technique can work well if a relatively obscure product becomes suddenly popular, but you do not have the time to determine its usefulness.

- Realize that not all social networking tools need to be evaluated with an enterprise or corporate application in mind. In some cases, a try it and see grassroots effort can reveal much more about the tool than a few people evaluating with a checklist.

Management may want more specific criteria for determining a Web 2.0 tool's fit for your organization. However, the process for evaluating social tools is not always the same as you would use for measuring enterprise collaboration tools. For example, you may need to be more experimental in your tests with social tools and use them under the radar before you can make a business case.

However, be aware that experimentation without management buy-in can be a risky strategy, and could cost you your job. Management may say, "We can't expend resources chasing the latest fad." Reasonable as that view of resource allocation may be, it may cause you to miss crucial conversations that might pop up during a pilot phase. Seizing the opportunity to engage with someone deeply may be more important than time and resource equations, but inappropriate or unsanctioned outreach may get you fired.

One short-lived collaboration tool was Google Wave, a personal communication and collaboration tool that didn't easily fit into any of the previous categories for social web content. Rather than tracking disparate pieces of conversations and content from many different Internet sources, objects known as "waves" contained a complete thread of multimedia messages. Because waves were XML-based, possibilities abounded for shared content.

Unfortunately, Google discontinued support for Google Wave in 2010. Many people felt that Google Wave reinvented email as we know it today. Gina Trapani wrote an entire user manual for it, donating the proceeds from the sales of the manual to charity. It is fascinating that even with a volunteer writing a manual for Google Wave out of her love for the product, it couldn't survive the cut.

User- and community-generated content

We may think of feedback as a modern web phenomenon, but in fact, we can find examples of feedback mechanisms throughout the history of publishing. Since began, there have always been methods for incorporating feedback into the final product. In his book *The Content Pool*[26], Alan Porter describes how Shakespeare's plays were modified each night based on audience feedback. The distributed work of the Oxford English Dictionary proposed in 1857 was to have contributors find quotations of a word's use by combing through all of English literature including popular newspapers, magazines, and then submit the quote slips to the editor of the dictionary. New findings of a word's use were brought into the next edition.

The key difference in today's publishing landscape is that the speed at which feedback and findings can come in has been accelerating, and you need to be aware of the ways that information can come in. Users also have taken over some of the publishing themselves as blogs and wikis and powerful searching have made it incredibly easy for any user's content to be noticed and used.

User-generated content is what you find in forums and mailing lists. It is likely to show up in a search for information and troubleshooting. User-generated content varies widely in quality and may be outdated quickly. Readers of user-generated content need to understand "Caveat Lector" – Reader Beware.

Community-generated content has a different quality bar. While a community may have a mailing list as its primary form of communication, it's likely the community is not using that mailing list to offer how-tos or detailed troubleshooting. Instead, a community, since it is defined with a common goal, may have a more formal means of content creation to achieve that goal.

Individual users are likely to be faster at posting informal, conversational responses to specific questions. But a community may have a more thought-out approach to the big picture of what content needs to be created. I am not just talking about written content. I'm also thinking of sites like Wordpress.tv, which was "seeded" with twenty or so professionally-created video tutorials, after which the community members' contributions were also accepted. While it may take a while to create that content, and it might not have a professional voice-over, it is good enough to help another community member learn a particular WordPress technique. And by leading with video offerings that gave an example of the quality level and expected pacing or chunking of the information, the community has a standard to meet.

Both types of content often offer "good enough" answers to questions or advice about the best way to proceed with a particular solution. Good enough is judged by the reader. One type of content generation is not more professional than another. Rather, the usefulness of the content vetted by a community is the criteria for judging its quality.

In my view, a set of community-generated content may be of higher quality than content created by individuals who do not share a common goal or want the same outcome. I would draw a distinction between user-generated content and community-generated content due to the motivations of a community being more aligned than a group labeled "users" who may be motivated by severely opposing aims. Chapter 3 explores community-generated content further.

B

Easter Seals Internet Public Discourse Policy

Easter Seals Internet Public Discourse Policy[1] SECTION III PART I-9 Approved by board: July 14, 2007

The Internet Public Discourse policy applies to Easter Seals headquarters and to Affiliates.

Easter Seals has always encouraged staff and volunteers to be champions on behalf of the organization by spreading the word about Easter Seals' work in providing life-changing solutions that help all people with disabilities have equal opportunities to live, learn, work and play.

The rapidly growing phenomenon of blogging, social networks and other forms of online electronic publishing are emerging as unprecedented opportunities for outreach, information-sharing and advocacy.

---

[1] This appendix copyright © Easter Seals. Used by permission.

Easter Seals encourages staff members and volunteers to use the Internet to blog and talk about our organization, our services and your work. Our goals are:

- To connect with and provide help and hope to children and adults with disabilities and the families who love them;

- To encourage support of Easter Seals' services and programs; and

- To share the expertise of Easter Seals' staff and volunteers.

Whether or not an Easter Seals staff member or volunteer chooses to create or participate in a blog or online community on their own time is his or her own decision. However, it is in Easter Seals' interest that staff and volunteers understand the responsibilities in discussing Easter Seals in the public square known as the World Wide Web.

Guidelines for Easter Seals Bloggers

1. Be Responsible. Blogs, wikis, photo-sharing and other forms of online dialogue (unless posted by authorized Easter Seals personnel) are individual interactions, not corporate communications. Easter Seals staff and volunteers are personally responsible for their posts.

2. Be Smart. A blog or community post is visible to the entire world. Remember that what you write will be public for a long time – be respectful to the company, employees, clients, corporate sponsors and competitors, and protect your privacy.

3. Identify Yourself. Authenticity and transparency are driving factors of the blogosphere. List your name and when relevant, role at Easter Seals, when you blog about Easter Sealsrelated topics.

4. Include a Disclaimer. If you blog or post to an online forum in an unofficial capacity, make it clear that you are speaking for yourself and not on behalf of Easter Seals. If your post has to do with your

work or subjects associated with Easter Seals, use a disclaimer such as this: "The postings on this site are my own and don't represent Easter Seals' positions, strategies or opinions." This is a good practice but does not exempt you from being held accountable for what you write.

5. Respect Privacy of Others. Don't publish or cite personal details and photographs about Easter Seals clients, employees, volunteers, corporate partners or vendors without their permission. Any disclosure of confidential information will be subject to the same Easter Seals personnel policies that apply to wrongful dissemination of information via email, conversations and written correspondence.

6. Write What You Know. You have a unique perspective on our organization based on your talents, skills and current responsibilities. Share your knowledge, your passions and your personality in your posts by writing about what you know. If you're interesting and authentic, you'll attract readers who understand your specialty and interests. Don't spread gossip, hearsay or assumptions.

7. Include Links. Find out who else is blogging about the same topic and cite them with a link or make a post on their blog. Links are what determine a blog's popularity rating on blog search engines like Technorati. It's also a way of connecting to the bigger conversation and reaching out to new audiences. Be sure to also link to easterseals.com

8. Be Respectful. It's okay to disagree with others but cutting down or insulting readers, employees, bosses or corporate sponsors and vendors is not. Respect your audience and don't use obscenities, personal insults, ethnic slurs or other disparaging language to express yourself.

9. Work Matters. Ensure that your blogging doesn't interfere with your work commitments. Discuss with your manager if you are

uncertain about the appropriateness of publishing during business hours.

10. Don't Tell Secrets. The nature of your job may provide you with access to confidential information regarding Easter Seals, Easter Seals beneficiaries, or fellow employees. Respect and maintain the confidentiality that has been entrusted to you. Don't divulge or discuss proprietary information, internal documents, personal details about other people or other confidential material.

Managing Content for Continuous Learning at Autodesk

When DITA Flows into a Social Web Platform

January 2011

By Geoffrey Bock and Dale Waldt

GILBANE GROUP
A DIVISION OF OUTSELL, INC.

Outsell's Gilbane Group
763 Massachusetts Avenue
Cambridge, MA 02139 USA
Tel: 617.497.9443
Fax: 617.497.5256
info@gilbane.com
http://gilbane.com

Managing Content for Continuous Learning at Autodesk: When DITA Flows into a Social Web Platform

Designing for Successive Generations

After a Pioneering Application

Founded in 1982, Autodesk, Inc., is a design software and services company, with customers in the architecture, engineering and construction, manufacturing, and digital media and entertainment industries. AutoCAD, its first notable product and a pioneering application in its day, ran on the first generation IBM Personal Computer and provided architects, designers, and engineers with an affordable tool to create detailed technical drawings. The company continues to exploit state-of-the-art computing systems and technologies. Today, Autodesk's varied products address all phases of the design processes, blending 2D and 3D design with 3D rendering, and extending results to encompass both Digital Prototyping and Building Information Modeling.

Key to its continuing appeal to successive generations of customers, Autodesk focuses on end-to-end experiences — ensuring that its products are easy-to-learn, even better-to-use, and popular among communities of technical professionals. Beyond simply publishing technical documentation, producing online help systems, and developing training tutorials, the company now seeks to promote customer loyalty through a continuous learning environment.

Supporting the Long Tail

Specifically, Autodesk has long realized that it cannot predict (and document) all the ways in which architects, designers, and engineers use its products. Over the years, these professionals have come to share their practical tips and best practices with one another through online communities. By capturing suggestions and sharing expertise among peers, these ad hoc groups provide the virtual venues where end users can learn how to apply design and modeling tools to a wide variety of specialized situations. Driven by customers and their different experiences, communities are the long tail of the Autodesk ecosystem, where the Internet combines the insights from many small groups into a major web presence.

Autodesk also senses a shift in how current generations of professionals, many of whom have been schooled in the digital age, expect to find information. Many prefer to consume content in short, pithy chunks, in the context of doing their work. Architects, designers, and engineers frequently find that interactive learning is more informative and relevant than paging through the published manuals and online help displays.

Changing Expectations

Autodesk thus faces the challenge of changing customer expectations. While it devotes substantial resources to publishing technical documentation, professionals today are relying less on this groomed and authenticated content. Continuing to invest in these conventional publishing solutions does not yield the anticipated results. Rather, Autodesk needs an environment that captures customers' varied learning styles, engages their intuitions and creativity, and interactively helps them solve their problems. Moreover, Autodesk needs to know when it is producing content that customers find useful.

With the widespread adoption of next generation social web platforms — environments designed to support wikis, tags, video, blogs, and user comments — there are now several new options. To maintain its position as a technology leader, Autodesk seeks to augment its groomed content, harness the user-generated content collected from online communities, make video an integral part of the ecosystem, and partner with customers to support the long tail of their experiences. Autodesk has thus embarked on an initiative to extend its conventional publishing practices, capture the insights from innovative users, incorporate video into online experiences, and produce compelling learning environments that drive future business.

Here's how.

Beyond Conventional Publishing

A Well-defined Content Infrastructure

To begin with, Autodesk already maintains a well-defined content infrastructure for developing and publishing technical documentation. Over the years, the company has continued to invest in content technologies, and is well advanced in its multi-year DITA adoption initiative. Product documentation groups are in the process of decomposing monolithic technical manuals into their component parts, tagging the granular components with Autodesk specializations of DITA, and storing them within a component content management (CCM) repository.

These investments are part of the cost of doing business – developing the authenticated content in a timely and efficient fashion, supporting the company's annual product release cycles, and publishing technical documentation in 19 languages. But adopting DITA, componentizing content delivery, and translating the same information into multiple languages does little to improve customer experience per se. Autodesk needs to not only produce technical information cheaper and faster, but also to make the overall customer experience better.

Focusing on the Customer Experience

By 2009, Autodesk began to realize that it needed to enhance its conventional publishing processes and extend them to support continuous learning activities. It was essential to help architects, designers, and engineers learn how to solve problems by focusing on three aspects of their experience:

- Improving customers' abilities to easily discover answers to questions about how Autodesk products work.

- Being able to distribute video and other types of rich media to make interactive instructions more informative and engaging.

- Leveraging the power of online communities by capturing and curating the contributions of innovative customers.

Furthermore, Autodesk needed to update its product content more frequently than once a year, target content delivery for problem solving, and better serve the more digitally savvy generations of customers. In addition to helping technical professionals better learn how to use its products, Autodesk needed to learn from them about what works.

Adopting a Social Web Platform

Fortunately, Autodesk can build upon its multiyear investment in DITA and its content infrastructure to begin to distribute content in new ways. The company can enrich its customer experience by adding a social web platform to its content infrastructure, and thus augment the ways in which it publishes information. This platform can easily deliver content segments around targeted topics, incorporate video into the interactive experiences, capture customer contributions from community interactions, and track results.

Specifically, Autodesk has added MindTouch to its content infrastructure to enhance content delivery and to support a continuous learning environment. This social web platform combines a wiki environment for web-based documentation with support for learning communities, social interactions, and content analytics. The wiki organizes content around the customer experience, and the ways in which practicing professionals typically look for answers to questions. For Autodesk customers, the social web takes over where conventional publishing leave off.

Business Benefits of a Social Web Presence

Wiki Delivery for Any Type of Content

How does a social web platform make the customer experience better?

Autodesk customers can browse sets of product-related topics, organized around problems and solutions. They are immersed in flexible and intuitive webs of information, rather than serial pages of documents and fixed indices of references.

As one expects from a contemporary web site, customers can click on relevant links, further drill into topic areas, find answers to questions, and rate results. The flow of content and the selection of relevant links change with the context. The wiki dynamically updates the multiple links to related topics. Notably in terms of the customer experience, content is not locked inside a predefined document or technical manual.

Whenever they want, customers can also search for information and filter results by various categories. The resulting items can include not only text and images, but also 3D models and video snippets. A social web platform can query, access, and render any type of digital content within an interactive environment. It ensures that Autodesk content is intelligently tagged and structured, so that it can be easily searched by customers and recognized by third party search engine crawlers. Text, audio, video, and other content types are managed, stored, and displayed within a unified and consistent environment.

Moreover, a social web platform supports "community pages" for collecting and sharing user-generated content on a wide variety of topics. Once authenticated by the site, both Autodesk customers and support engineers from across the ecosystem can publish "tips and tricks" recommendations, blog on new topics, comment on posts, reference code snippets, or add links to topics covered elsewhere on the web. The platform manages the security in a granular fashion so that only users with predetermined rights can perform specific actions.

Finally, MindTouch tracks how customers use content. It logs when and how customers access content. Featuring integrated content analytics capabilities, this social web platform identifies both the sequence of links within the site and the references to external Internet connections. It can organize results to determine the popularity of particular topics, the frequency of search terms, the relationships among search terms and content items, and the sources of Internet connections. With these kinds of tracking capabilities, Autodesk can easily determine which topics are most popular and relevant, and plan its content development activities accordingly.

Managing the Content Flow and Metadata

Consequently, Autodesk leverages its existing content infrastructure to support its social web presence. The company manages both audio and video content within its own content infrastructure, without relying on third party resources.

Already familiar with an XML authoring environment, technical writers at Autodesk develop, tag, and store DITA content as part of their conventional publishing processes. They not only continue to produce technical information to support the annual product release cycles, but also are able to republish content in an entirely different context.

Significantly, the content at Autodesk is sufficiently granular and enriched with relevant metadata to provide the key connections for the social web experience. Consistent metadata, defined by DITA and other XML tags, creates the semantic links for content to flow between the conventional publishing system and the social web environment. This is an example of multi-purposing, where content initially developed for one purpose can be reused for purposes beyond those that are initially envisioned.

For instance, documentation teams already structure and tag online help topics in terms of "concept," "procedure," and "reference," and then publish the results as separate sections within the online system. This content can also be automatically incorporated into the social web platform to produce another customer experience. Within this platform, topics are concatenated and published as a single wiki page with embedded links to the more detailed information. Customers can now offer their own comments and best practice suggestions about the published information.

Furthermore, tagging within a social web platform is flexible and adaptive. The wiki environment does not depend on a predefined hierarchy. Certainly, DITA serves as a starting point. For example, DITA tags that define the product names and product versions can be automatically turned into a tag cloud when delivered through a social web platform. But other important criteria, such as familiar words and phrases that customers frequently use when searching for items, can also be added to the content as additional tags. These customer contributions can help improve content search, retrieval, and delivery. In short, managing the metadata is an important way to drive the customer experience and refine the delivery of content as a social, customer-driven experience.

Two Perspectives on Continuous Learning

In fact, the social web platform supports continuous learning experience from two perspectives. Digitally savvy customers have an interactive venue where they can better learn how to use Autodesk products. At the same time, Autodesk can rely on this platform to learn about its customers.

With a social web platform, Autodesk can easily capture customer contributions from community interactions, identify those that are most popular or that highlight unique applications, and incorporate the best of this user-generated content into the published content. Autodesk can easily close the feedback loop, and continuously learn about its customers.

From Content Developers to Content Curators

New tasks lead to new roles. With a social web platform, technical writers and editors not only document product capabilities, but also help customers learn about using features and functions, and guide their discovery of relevant information.

In addition to supporting conventional publishing activities, writers and editors at Autodesk also pay attention to what works for customers. Some team members are shifting roles from simply developing technical information to curating content. In addition to being responsible for documenting a topic area, they are also concerned about how information sets are organized, accessed, and used.

As content curators, team members track what technical professionals do, identify the popular topics, and maintain the categories used to organize the content on the wiki pages. Content curators not only develop and source the published information snippets. They also define new metadata categories to provide additional ways for navigating the wiki and for molding the customer experience.

Finally, content curators can track the customer contributions across the multiple online communities within Autodesk ecosystem. When developing content for a forthcoming release, they can identify the most popular and the most relevant customer contributions published within the wiki environment. With a social web platform incorporated into its content infrastructure, Autodesk can better determine what works and plan its content development investments accordingly.

Smart Content Insights

With its investment in DITA and a component content management system, Autodesk demonstrates how a social web platform can embrace and extend the capabilities of a well-defined content infrastructure, and deliver smart content to foster continuous learning. The Autodesk experience highlights the business benefits of the social web for interactive problem solving.

Digitally savvy professionals can quickly explore topic areas, learn about new product features, and find answers to questions. The published information combines both the authenticated information produced by technical documentation teams, and the content customers generate through community-wide discussions.

Significantly, technical documentation teams have new and expanded roles. Content developers, familiar with the details of product capabilities, become the content curators who track how customers use the information, identify innovative solutions, continuously curate and tag the additional content, and help to quickly spread the word about what works.

Content Delivery. For Autodesk, content delivery and the customer experience are the starting points for application design. With its social web platform, Autodesk can organize and deliver any type of content, including video segments, as part of an integrated and seamless experience. In addition, Autodesk can better understand how customers use the published content, both by tracking what's popular and by collecting and refining the user-generated comments.

Content Enrichment. Content at Autodesk is not only structured and granular. It gets smarter over time. Specifically, Autodesk builds on its existing investment in DITA by defining the granularity of content at the appropriate level and by tagging it with semantic metadata. Autodesk continues to enrich the content in light of customer experiences by adding additional metadata elements. Using this metadata within a social web platform, Autodesk can tailor content delivery around targeted business purposes.

Content Creation. Content delivered on the social web is part of the overall content infrastructure. Content development with DITA is the beginning of the process. Content developers create the granular content components where they add relevant metadata as they index their results. By tagging the information in a systematic fashion, the technical product information flows into the social web platform.

Furthermore, among the technical documentation team, content developers become content curators. In addition to describing the details of product features, they are also concerned about creating the compelling customer experiences, and maintaining them over time in light of what digitally savvy professionals do. Content curators continuously enrich the content for the social web to make it smarter.

Glossary

Agile development

A software development methodology that develops software rapidly with multiple tight iterations, or development cycles (2–3 weeks). Each iteration includes design, development, test, and documentation of a subset of the product's features. Each iteration should result in a potentially deliverable product.

Atom

An XML standard, similar to RSS, used for syndication. (See also *Syndication.*)

Barcamp

A user-focused conference where the content is provided by participants. The "bar" in the name refers not to the place it is held but to the hacker term "foobar," as an opposing camp for Tim O'Reilly's Foo Camp – an invitation-only event. Barcamp includes people with similar interests in open source and technology.

Blog

An online website where entries are typically listed in reverse chronological order. The term comes from a contraction of the term "weblog."

Blogosphere

A term that describes the interconnected world of blogs, bloggers, and their culture.

Book sprint

An event for collaborative authoring that typically starts with an outline and results in a book by the end of a set period of time, usually about a week. Usually set in one location at one time, but remote collaborators can also participate.

Bookmarking

Saving a web address for later retrieval and reading, similar to placing a piece of paper between pages to mark that spot for reading a book. (See also *Social bookmarking.*)

Bot, robot

Software that automatically searches the web for specific content on command.

Community

A group of people with similar interests or goals. An online community typically has a central location online to share these interests.

Community manager

An emerging role, still being defined, that involves communication with the community, encouraging community members, advocating for the community, and promoting awareness.

Content curator

A curator is typically the custodian of a collection of art or historic objects; for content curation the person's role is to manage, administer, and generally take care of content.

Creative Commons license

A set of copyright licenses that allow a variety of options for sharing content. They give authors the ability to authorize different uses of the material while retaining control.

Crowdsourcing

A method for distributing a task among a group of people who are asked to volunteer their efforts to accomplish the task.

Similar to outsourcing, but instead gives the work to a large group of people with no formal compensation other than recognition.

Curation
Organizing and maintaining a collection.

DITA (Darwin Information Typing Architecture)
An XML-based extensible architecture for creating technical documentation. DITA has two defining characteristics: a topic-based architecture and a customization facility called "specialization" that lets you extend the grammar while maintaining interoperability with others who use DITA. The DITA standard is maintained by OASIS Open.

DITA map
A DITA map structure organizes DITA topics into information deliverables.

DocBook
An XML-based standard for creating technical documentation. Widely used in the open source community and broadly supported by open source tools. This book was written and produced in DocBook. The DocBook standard is maintained by OASIS Open

Enabler
A social media role. An enabler of conversation understands the underlying concepts of a product or service well enough to help others have an intelligent conversation about it.

Flame wars
A heated argument online that results in back-and-forth discussion including insults and angry remarks.

FLOSS Manuals
Both a community and web-based collaboration tool created in 2007 by Adam Hyde for the purpose of providing free docu-

mentation for free software. FLOSS in the name is an acronym for Free Libre Open Source Software.

Folksonomy

A classification system using tags created by members of an online community or everyday folks. Plays off the word "taxonomy," which is the formal practice and science of classifying items.

Forum

A message board system used online for public or member-only discussion and debate on topics.

FrameMaker

A word processing and publishing software product from Adobe, often used in technical documentation for large documents.

Gift economy

A system where effort is bartered in exchange for purely altruistic gains.

Going viral

A phenomenon where a video, image, or text becomes extremely popular on the Internet primarily by word of mouth.

Groundswell

A social trend in which people use technologies to get the things they need from each other rather than from traditional institutions like corporations.

Haptic

A user interface that uses sensory feedback. Examples include touchscreen keyboards and the Nintendo Wii controller, which causes a "bump" sensation when moving the cursor over a button.

Hashtag

Words or phrases prefixed with a hash symbol (#) used on Twitter to identify content for later searches, grouping, and retrieval. An example is the #stc12 hashtag used to indicate that a Twitter update was about the STC Summit in 2012.

Instant Messaging (IM)

A form of online chat that enables real-time messages sent back and forth between two or more people. Typically an IM client is used to enable the chat. Some clients support text messages only while others enable video and audio streaming in addition to text.

Instigator

A social media role. An instigator starts a discussion, perhaps by offering a controversial opinion or a highly debatable strategic choice.

Internet Relay Chat (IRC)

A chat network protocol on the Internet that enables synchronous communication by typing text in hosted channels.

Mashup

An online application or tool that combines data or content from multiple sources to give a new perspective or service.

MediaWiki

A web-based wiki software application. Its most well-known use is for the online encyclopedia Wikipedia.

Microblog, microblogging

Brief text-based entries that are smaller than normal blog posts. As an example, it describes Twitter posts with their 140-character per post limitation.

One Laptop per Child (OLPC)

A US-based non-profit organization whose mission is to provide affordable, rugged education devices to children in developing countries.

Open source

A method for distributing software that guarantees anyone who uses that software the right to use, modify, and re-distribute the software and its source code freely. Usually, open source software must be re-distributed under the same terms under which it was originally distributed. Typically open source projects are built and maintained by volunteers. Well-known examples of open source projects include the Apache web server and the Firefox web browser.

PDF (Portable Document Format)

A file format developed by Adobe that preserves all printed page formatting for electronic viewing and distribution.

Podcast

An audio recording that is broadcast on the Internet. Often uses RSS for subscribing to updates.

Round tripping

The act of taking content from one application into another and then back again to the original application.

RSS

RSS stands for Really Simple Syndication. (See also *Syndication*.)

Screen scraping

Extracting data, typically HTML, from a web page using a program.

Screencasting

A recording of your screen while you are doing a computer task on-screen that may include narrative instructions or explanation of the software being used.

Search Engine Optimization (SEO)

The process of improving the search engine ranking of content and increasing the number of click-throughs to a website. Typically involves optimizing the use of keywords and eliminating barriers to automated indexing activities performed by search engines.

Social bookmarking

The practice of sharing web addresses for storage and retrieval with others. First coined by the site http://delicious.com, which incorporates tagging and a folksonomy for applying metadata to bookmarks.

Social media

Using media for sharing and social interactions as well as spreading information. Typically uses web-based, scalable, interactive techniques. Often contrasted to traditional broadcast media because it enables more publishers, authors, voices, and channels to be heard.

Social networking

The practice of connecting and communicating with people who share interests or common goals. Social networking sites enable this communication and sharing online.

Social relevance

Refers to aligning and enhancing a brand or product's strategy by increasing its relevance to typical social web audiences, influencers, mobile web users, and others with specific tactics and goals in mind.

Social shopping

A method of offering social interactions or communications revolving around e-commerce websites using techniques such as product reviews or forums embedded within the shopping experience.

Social tagging

Assigning keywords for a blog entry, link, photo, or video, and then sharing them with the community online or a select network of other users of the website. (See also *Tagging*.)

Social weather

A multi-dimensional visualization of the relative attitude of a community or group.

Social web

Describes people applying Internet technology in a community or socially interactive manner using reputation, identity, presence, relationships, grouping, communication, and sharing to accomplish tasks or projects.

Syndication

Supplying content such as text, images, video, or audio, for consumption, reuse, and integration with other material via a subscription feed. RSS and Atom are XML standards that support syndication.

Tag cloud

A visual method of displaying tags (keywords). Tag clouds use a larger font size or bold to indicate which tags are associated with the largest number of references or most recently accessed references.

Tagging

The act of assigning a keyword to a particular item such as a blog entry, link, photo, or video.

Trackback

Offers a method for tracking links to content by sending a notification to another site, typically a blog, when one entry links to another entry. Often, snippets of text from the blog content are included with the trackback link and notification.

Tweet

A single post on Twitter, or when used as a verb, the act of posting an update to Twitter. (See also *Twitter*.)

Tweetup

A real-world meeting enabled by using Twitter to invite, communicate with, and respond to participants about the event. Substitutes Tweet for Meet in the term "meet up."

Twitter

A web application that enables you to post 140-character micro-entries from the web or from a mobile phone answering the question, "What are you doing?"

Unconference

A user-driven conference facilitated by organizers, but with the content of the conference shaped by the attendees around a theme or purpose.

Vlog, video log

A word play on the term "blog," it describes video publications that are syndicated and subscribed to similar to a blog.

Web 2.0

The second version of the World Wide Web where interaction and collaboration takes precedence over static data.

Whuffie

The ephemeral, reputation-based currency invented by Cory Doctorow in his sci-fi novel, *Down and Out in the Magic Kingdom*[7].

Wiki

A collaborative website where pages can be edited and added to by contributors.

Wiki pattern

One of a set of known solutions to known problems related to wikis. Based on the concept of architectural patterns. Coined by Stewart Mader and described in his book, *Wikipatterns*[21].

Wikislice

A cross-section of selected wiki pages collected together to form a new deliverable.

Wikitext

A markup language used to create the web pages that make up a wiki. Often it is a simplified, ASCII-based alternative to HTML coding.

Word Press

An open source blog publishing application.

XHTML

Acronym for eXtensible HyperText Markup Language. A W3C standard that applies XML syntax to HTML markup.

XML

Acronym for eXtensible Markup Language. A W3C standard for building markup languages. DITA, DocBook, XHTML, and dozens of other standards are based on the XML standard. XML imposes structure on a document through elements, which surround content (for example, `<para>`some content.`</para>`), and attributes (like `lang` in `<para lang="fr">`), which provide additional information.

XSLT

Acronym for eXtensible Stylesheet Language Transformations. A W3C standard, XSLT is an XML-based language for transforming XML files. Typically, XSLT is used to process an XML instance, generating another XML instance. For example, the DocBook and DITA stylesheets use XSLT to transform DocBook or DITA instances into ouput forms like HTML, PDF (through intermediate processing), and Javahelp.

Bibliography

Books

[1] Ament, Kurt. 2002. *Single Sourcing: Building Modular Documentation.* William Andrew. ISBN: 0815514913.

[2] Anderson, Chris. 1991. *The Long Tail: Why the Future of Business is Selling Less of More.* Hyperion. ISBN: 978-1-4013-0966-4.

[3] Barefoot, Darren and Julie Szabo. 2009. *Friends with Benefits.* No Starch Press. ISBN: 978-1593271992. Updated version of the privately published eBook, *Getting to First Base: A Social Media Marketing Playbook.*

[4] Barrett, Edward. 1991. *The Society of Text: Hypertext, Hypermedia, and the Social Construction of Information.* The MIT Press. ISBN: 026252161X.

[5] Brafman, Ori. 2006. *The Starfish and The Spider: The Unstoppable Power of Leaderless Organizations.* Portfolio Hardcover. ISBN: 1591841437.

[6] Carroll, John M. 1990. *The Nurnberg Funnel: Designing Minimalist Instruction for Practical Computer Skill.* The MIT Press. ISBN: 978-0262031639.

[7] Doctorow, Cory. 2003. *Down and Out in the Magic Kingdom.* Tor Books. ISBN: 978-0765309532.

[8] Dreyfus, Hubert L. and Stuart E. Dreyfus. 1988. *Mind over Machine: The Power of Human Intuition and Expertise in the Era of the Computer.* Free Press. ISBN: 978-0029080610.

[9] Ferris, Timothy. 2007. *The 4-Hour Workweek: Escape 9-5, Live Anywhere, and Join the New Rich*. Crown Publishing. ISBN: 0307353133.

[10] Friedman, Thomas. 2007. *The World Is Flat 3.0: A Brief History of the Twenty-first Century*. Picador. ISBN: 0312425074.

[11] Gladwell, Malcomb. 2001. *The Tipping Point: How Little Things Can Make a Big Difference*. Back Bay Books. ISBN: 0316346624.

[12] Halvorson, Kristina and Melissa Rauch. 2012. *Content Strategy for the Web*. 2nd edition. New Riders Press. ISBN: 978-0-321-80830-4.

[13] Holovaty, Adrian and Jacob Kaplan-Moss. 2009. *The Definitive Guide to Django: Web Development Done Right*. 2nd edition. APress. ISBN: 978-1430219361. This web "preview" version has per-paragraph commenting: http://www.djangobook.com/en/2.0/-chapter01/.

[14] Hunt, Tara. 2009. *The Whuffie Factor: Using the Power of Social Networks to Build Your Business*. Crown Business. ISBN: 978-0307409508.

[15] Jones, Colleen. 2011. *Clout: The Art and Science of Influential Web Content*. New Riders Press. ISBN: 978-0-321-73301-6.

[16] Keen, Andrew. 2007. *The Cult of the Amateur: How Today's Internet is Killing Our Culture*. Broadway Business. ISBN: 0385520808.

[17] Klobas, Jane. 2006. *Wikis: Tools for Information Work and Collaboration*. Chandos Publishing (Oxford) Ltd. ISBN: 1843341786.

[18] Levine, Rick, Christopher Locke, Doc Searls, and David Weinberger. 1999. *The Cluetrain Manifesto: 10th Anniversary Edition*. Basic Books. ISBN: 978-0465024094. Available online at: http://-www.cluetrain.com/.

[19] Li, Charlene and Josh Bernoff. 2008. *Groundswell: Winning in a World Transformed by Social Technologies*. Harvard Business School Press. ISBN: 1422125009.

[20] Maddox, Sarah. 2012. *Confluence, Tech Comm, Chocolate: A wiki as platform extraordinaire for technical communication.* XML Press. ISBN: 978-0-937434-00-7.

[21] Mader, Stewart. 2007. *Wikipatterns.* Wiley. ISBN: 0470223626.

[22] OpenStack. 2012. *OpenStack Compute Starter Guide.* OpenStack. http://docs.openstack.org/essex/openstack-compute/starter/content/index.html.

[23] O'Reilly, Tim and Sarah Milstein. 2009. *The Twitter Book.* O'Reilly Media. ISBN: 0596802811.

[24] Peterson, Eric T. 2004. *Web Analytics Demystified.* Celilo Group Media. ISBN: 0-9743584-2-8. Available as free download from: http://www.webanalyticsdemystified.com.

[25] Porter, Alan J. 2010. *WIKI: Grow Your Own for Fun and Profit.* XML Press. ISBN: 978-0-9822191-2-6.

[26] Porter, Alan J. 2012. *The Content Pool: Leveraging Your Company's Largest Hidden Asset.* XML Press. ISBN: 978-0-937434-01-4.

[27] Redish, Janice (Ginny). 2007. *Letting Go of the Words: Writing Web Content that Works.* Morgan Kaufmann. ISBN: 0123694868.

[28] Rheingold, Howard. 2002. *Smart Mobs: The Next Social Revolution.* Basic Books. ISBN: 0738206083.

[29] Rockley, Ann and Charles Cooper. 2012. *Managing Enterprise Content: A Unified Content Strategy.* 2nd edition. New Riders Press. ISBN: 978-0-321-81536-1.

[30] Scoble, Robert and Shel Israel. 2006. *Naked Conversations: How Blogs are Changing the Way Businesses Talk with Customers.* Wiley. ISBN: 047174719X.

[31] Shirky, Clay. 2009. *Here Comes Everybody: The Power of Organizing Without Organizations.* Penguin (Non-Classics). ISBN: 0143114948.

[32] Shirky, Clay. 2011. *Cognitive Surplus: Creativity and Generosity in a Connected Age.* Penguin Books. ISBN: 978-0141041605.

[33] Sun Technical Publications. 2009. *Read Me First! A Style Guide for the Computer Industry.* 3rd edition. Prentice Hall. ISBN: 978-0137058266.

[34] Surowiecki, James. 2005. *The Wisdom of Crowds.* Anchor. ISBN: 0385721706.

[35] Tapscott, Don and Anthony D. Williams. 2008. *Wikinomics: How Mass Collaboration Changes Everything.* Portfolio Hardcover. ISBN: 1591841933.

[36] Wurman, Richard Saul. 2000. *Information Anxiety 2.* Que. ISBN: 978-0789724106. Follow up to his 1989 classic, *Information Anxiety.*

[37] Zelenka, Anne Truitt. 2008. *Connect!: A Guide to a New Way of Working from GigaOM's Web Worker Daily.* Wiley. ISBN: 0470223987.

Articles, websites, blogs

[38] @twitter. "#numbers." March 14, 2011. http://blog.twitter.com/2011/-03/numbers.html . This blog post shows a variety of growth figures for twitter as of the publication date.

[39] Abel, Scott and Stewart Mader. "Why Businesses Don't Collaborate." http://www.scribd.com/doc/16336782/Why-Businesses-Dont-Collaborate-Meeting-Management-Group-Input-and-Wiki-Usage-Survey-Results.

[40] Aguiar, Ademar, Paulo Merson, and Uri Dekel. "Using Structured Wikis in Software Engineering." *WikiSym '08 Proceedings of the 4th International Symposium on Wikis.* (Porto, Portugal, September 8-10, 2008). ACM. ISBN: 978-1-60558-128-6. https://-dl.acm.org/citation.cfm?id=1822308.

[41] Anderson, Chris. "Free! Why $0.00 Is the Future of Business." http://-www.wired.com/techbiz/it/magazine/16-03/ff_free.

[42] Bailie, Rahel. "Managing Online Communities: The Next Big Idea for Communications Professionals?" http://www.enewsbuilder.net/techcommanager/e_article001389770.cfm?x=b11,0,w.

[43] Baker, Mark. "I am a content strategist." http://everypageispageone.com/2012/05/04/i-am-a-content-strategist/.

[44] Bernoff, Josh. "Social Technographics: Conversationalists get onto the ladder." http://forrester.typepad.com/groundswell/2010/01/conversationalists-get-onto-the-ladder.html.

[45] Blankenhorn, Dana. "Why open source documentation lags." http://www.zdnet.com/blog/open-source/why-open-source-documentation-lags/6484.

[46] Bleiel, Nicky. "Convergence Technical Communication: Strategies for Incorporating Web 2.0." http://www.thecontentwrangler.com/article/convergence_technical_communication_strategies_for_incorporating_web_20/.

[47] Bock, Geoffrey and Steve Paxhia. "Collaboration and Social Media—2008." The Gilbane Group. June 9, 2008. http://gilbane.com/Research-Reports/Gilbane-Social-Computing-Report-June-08.pdf.

[48] boyd, danah. "Social Media is Here to Stay... Now What?" http://www.danah.org/papers/talks/MSRTechFest2009.html. Transcript of a talk danah gave at Microsoft Research Tech Fest, Redmond, WA, 26 February 2009.

[49] Bringhurst, Bob. "Best InDesign Links of May." http://blogs.adobe.com/indesigndocs/2009/05/best_indesign_links_of_may.html.

[50] Community Roundtable. "The 2010 State of Community Management: Best Practices from Community Practitioners." http://communityroundtable.com/socm-2010/.

[51] Dagenais, Barthélémy and Martin P. Robillard. "Creating and evolving developer documentation: understanding the decisions of open source contributors." *FSE '10 Proceedings of the eighteenth ACM SIGSOFT international symposium on Foundations of software*

engineering. (Santa Fe, NM, November 7-11, 2010). ACM. ISBN: 978-1-60558-791-2.

[52] GNU Project. "What is free software?" http://gnu.org/philosophy/free-sw.html.

[53] Gentle, Anne. "The 'Quick' Web for Technical Documentation." *STC Intercom.* September/October 2007). pp. 16-19. http://archive.stc.org/intercom/PDFs/2007/20070910_16-19.pdf (STC members only).

[54] Gentle, Anne. "Can online help show 'read wear'?" http://justwrite-click.com/2008/05/28/can-online-help-show-read-wear/.

[55] Gunderloy, Mike. "The wiki-fication of MSDN." http://searchwindevelopment.techtarget.com/news/article/-0,289142,sid8_gci1211391,00.html.

[56] Halavais, A. "Blogs and the 'social weather.'" Paper presented at *Internet Research 3.0.* (Maastricht, The Netherlands, October, 2002). pp. 3-9.

[57] Halvorson, Kristina. "The Discipline of Content Strategy." http://searchwindevelopment.techtarget.com/news/article/-0,289142,sid8_gci1211391,00.html.

[58] Happe, Rachel. "Social Media is not Community." http://www.thesocialorganization.com/2008/07/social-media-is-not-community.html.

[59] Heil, Bill and Mikolaj Piskorski. "Twitter – New Research: Men Follow Men and Nobody Tweets." http://www.iq.harvard.edu/blog/netgov/2009/06/hbs_research_twitter_oligarchy.html.

[60] Hill, William C., et al. "Edit wear and read wear." *Proceedings of CHI '92, the SIGCHI Conference on Human Factors in Computing Systems.* (Monterey, CA, May 3-7, 1992). pp. 3-9.

[61] Holland, Anne. "How to Calculate a Blog's Reach & Influence – More Complex Than You Think." http://www.marketingsherpa.com/-article.html?ident=30044.

[62] Howe, Jeff. "The Rise of Crowdsourcing." http://www.wired.com/-wired/archive/14.06/crowds.html.

[63] Huba, Jackie. "Keeping up with the social media fire hose." http://-www.churchofthecustomer.com/blog/2008/04/keeping-up-with.html.

[64] Johnson, Tom. "A Web 2.0 Documentation Idea Gone Wrong." http://www.idratherbewriting.com/2008/02/06/a-web-20-document-ation-idea-gone-wrong/.

[65] Kanter, Beth. "How Much Time Does It Take To Do Social Media?" http://beth.typepad.com/beths_blog/2008/10/how-much-time-d.html.

[66] Kollock, Peter. "The Economies of Online Collaboration: Gifts and Public Goods in Cyberspace." In Kollock, Peter and Marc Smith. *Communities in Cyberspace.* London: Routledge. 1999. ISBN: 978-0415191401. pp. 220-239. Available online at: http://www.con-nectedaction.net/wp-content/uploads/2009/05/2001-peter-kollock-economies-of-online-cooperation.htm.

[67] Maddox, Sarah. "The agile technical writer." http://ffeathers.word-press.com/2008/01/20/the-agile-technical-writer/.

[68] Maddox, Sarah. "Twitter as a medium for release notes." http://ffeath-ers.wordpress.com/2009/06/08/twitter-as-a-medium-for-release-notes/.

[69] McCandless, David. "Cognitive Surplus Visualized." http://www.in-formationisbeautiful.net/2010/cognitive-surplus-visualized/.

[70] Miller, Harry. "The IM Model of Tech Writing." http://-blogs.msdn.com/harrymiller/archive/2007/07/16/the-im-model-of-tech-writing.aspx.

[71] Nielsen, Jakob. "Participation inequality: Encouraging more users to participate." http://www.useit.com/alertbox/participation_inequal-ity.html.

[72] Oram, Andy. "Rethinking Community Documentation." http://www.on-lamp.com/pub/a/onlamp/2006/07/06/rethinking-community-documentation.html.

[73] Oram, Andy. "Why Do People Write Free Documentation? Results of a Survey." http://www.onlamp.com/pub/a/onlamp/2007/06/14/-why-do-people-write-free-documentation-results-of-a-survey.html.

[74] Owyang, Jeremiah. "Defining the term: 'Online Community.'" http://-www.web-strategist.com/blog/2007/12/28/defining-the-term-community/.

[75] O'Keefe, Sarah. "Friend or Foe? Web 2.0 in Technical Communication." http://www.scriptorium.com/whitepapers/web2/.

[76] Porter, Alan. "Why publishing is no longer the last step in the process." Slide set available at: http://www.slideshare.net/webworks/why-publishing-is-no-longer-the-last-step-2513597.

[77] Raymond, Eric S. "Goodbye 'free software'; hello, 'open source.'" http://www.catb.org/~esr/open-source.html. This 1998 article is the first to use the term "open source" in the software community.

[78] Saleem, Muhammad. "Writing for the Social Media Everyman." http://www.copyblogger.com/writing-for-social-media/.

[79] Self, Tony. "What if readers can't read?" http://www.hyperwrite.com/-Articles/showarticle.aspx?id=84.

[80] Shirky, Clay. "Group as User: Flaming and the Design of Social Software." In Spolsky, Joel, ed. 2005. *The Best Software Writing I.* Berkeley, CA: Apress. ISBN: 978-1-59059-500-8. pp. 211-221. Available online at: http://shirky.com/writings/group_user.html.

[81] Sierra, Kathy. "How to get users to RTFM." http://headrush.type-pad.com/creating_passionate_users/2006/09/-how_to_get_user.html.

[82] Spool, Jared. "Where Did Technical Writing Go?" http://www.uie.com/-brainsparks/2007/05/16/where-did-technical-writing-go/.

[83] Swartz, Aaron. "Who Writes Wikipedia." http://www.aaronsw.com/-weblog/whowriteswikipedia.

[84] Swisher, Janet. "Embrace the 'Un' – When the Community Runs the Event." *STC Intercom*. January 2009. http://justwriteclick.com/-2009/01/17/embrace-the-un/.

[85] Technorati. "State of the Blogosphere, February 2006." http://www.sifry.com/alerts/archives/000419.html. For more recent data, see: http://technorati.com/blogging/state-of-the-blogosphere/.

[86] Tenney, Alvan. A. "The scientific analysis of the press." *Independent* 73 (1912): 895-898.

[87] VanFossen, Lorelle. "When is the Best Time and Day to Publish a Blog Post?" March 14, 2011. http://lorelle.wordpress.com/2008/12/03/-when-is-the-best-time-and-day-to-publish-a-blog-post/. Updated each year with new information.

[88] Vernon, Amy. "Creating a library of FLOSS Manuals." http://www.networkworld.com/community/node/61301.

[89] Wikinews. "Jack Herrick, wikiHow founder interviewed by Wikinews." http://en.wikinews.org/wiki/Jack_Herrick,_WikiHow_founder_interviewed_by_Wikinews.

[90] Zelenka, Anne. "From the Information Age to the Connected Age." http://gigaom.com/2007/10/06/from-the-information-age-to-the-connected-age/.

Index

About XML Press

XML Press (http://xmlpress.net) was founded in 2008 to publish content that helps technical communicators be more effective. Our publications support managers, social media practitioners, technical communicators, content strategists, and the engineers who support their efforts.

Our publications are available through most retailers, and discounted pricing is available for volume purchases for business, educational, or promotional use.

For more information, visit our website at http://xmlpress.net, send email to orders@xmlpress.net, or call us at (970) 231-3624.

CPSIA information can be obtained at www.ICGtesting.com
Printed in the USA
BVOW11s1653270814

364392BV00010B/129/P